Aspects of the Structure, Cytochemistry and Germination of the Pollen of Rye
(*Secale cereale* L.)

Aspects of the Structure, Cytochemistry and Germination of the Pollen of Rye (*Secale cereale* L.)

J. HESLOP-HARRISON

Welsh Plant Breeding Station, Plas Gogerddan, Aberystwyth SY23 3EB

Reprinted from
Supplement 1 (pp. 1–47 with 18 plates) to
Annals of Botany, Volume 44, 1979

1980
Published for
The Annals of Botany Company
by
Academic Press
A Subsidiary of Harcourt Brace Jovanovich, Publishers
London   New York   Toronto   Sydney   San Francisco

ACADEMIC PRESS INC. (LONDON) LIMITED
24/28 Oval Road
LONDON NW1
(Registered number 598514)

US edition published by
ACADEMIC PRESS INC.
111 Fifth Avenue
New York
New York 10003

All rights reserved
No part of this book may be reproduced in any form by photostat, microfilm or any other means
without written permission from the publishers

© 1980 Annals of Botany Company

British Library Cataloguing in Publication data.

Heslop-Harrison, John
Aspects of the structure, cytochemistry, and germination of the pollen of the rye. –
('Annals of botany' supplements; 1).
1. Rye 2. Pollen 3. Germination
I. Title II. Series
584'.93   QK495.G74   79-41655
ISBN 0-12-344950-2

PREFACE

Important although the grasses are in human affairs, many basic features of their biology remain to be investigated. This is true for aspects of the reproductive physiology of the group, and notably for those connected with pollen, the pollen–stigma interaction and pollen-tube growth. This short memoir reports some of the results of my own recent work on the structure and germination of the pollen of rye and other grass species. The treatment is itself in the nature of an experiment, combining as it does something of the character of both research paper and review. This is the format envisaged by the Annals of Botany Company in launching this new series of supplements. For me it has offered the opportunity not only to report new findings but to relate them to those of other workers in rather more detail than is usually possible within the confines of the normal research paper. I am of course thoroughly well aware of the great lacunae in our knowledge of many of the structures and processes I describe, and am sensitive of the fact that the treatment I have been able to give many of the topics falls very far short of what one might have wished. However, perhaps the very act of juxtaposing cytological and physiological aspects that are not often considered together may stimulate further enquiry; and if so I shall be satisfied.

J. HESLOP-HARRISON

LIST OF CONTENTS

PREFACE	3
ABSTRACT	5
INTRODUCTION	5
The grass pollen grain: a summary of structural features	5
Hydration and germination	8
MATERIALS AND METHODS	12
OBSERVATIONS AND EXPERIMENTAL RESULTS	18
The pollen wall	18
Cytoplasm and organelles in ungerminated pollen	22
Hydration and germination	23
DISCUSSION	33
ACKNOWLEDGEMENTS	39
LITERATURE CITED	40

ABSTRACT

The paper provides a detailed account of certain ultrastructural features of the pollen of *Secale cereale* L. (Gramineae), supplemented by cytochemical data for the wall and components of the vegetative cell. In the ungerminated grain the membranes are ill-defined, but they become reorganized within 2–3 min from the beginning of hydration. A scheme is developed to account for the behaviour of the grain during imbibition, based on the assumption that the membranes do not initially form an osmotic barrier. The Zwischenkörper, which is composed of gel-forming pectins, is shown to play an important part in germination. By the use of intine 'ghosts' prepared by the removal of the exine and protoplast, the changes in the intine at the apertural site have been followed during the earliest stages of germination. The wall of the emerging tube tip has a cellulosic component formed of short, randomly oriented microfibrils; behind this, the microfibrils are oriented in a circumferential manner, and an inner callosic wall is deposited. The precursors of the wall appear to be contributed by a population of polysaccharide-containing particles ('P-particles') present in the ungerminated grain. Bodies of a similar nature arise in association with the starch of the grain, and later also from the intine, which is ultimately consumed during the growth of the tube. There is no conspicuous dictyosome activity in the newly emerging tube, which does not show the type of tip zonation described from other monocotyledonous families. The emerging tip secretes a pectinase and certain other enzymes; the source of this secretion may be a population of vesicles present in the cytoplasm of the tube tip.

Earlier work on the structure, cytochemistry, hydration and germination of the grass pollen grain is reviewed.

Key words: *Secale cereale* L., rye, pollen grain, structure, cytochemistry, germination, grass pollen.

INTRODUCTION

The grass pollen grain: a summary of structural features

The wall

The pollen wall of the grasses shows the familiar differentiation into an outer exine and an inner intine. The main features of the 'exine' were made clear by the work of Erdtman (1943, 1952) using optical microscopy. The non-apertural exine was shown to be tectate, the tectum being raised upon short bacula from a foot layer which overlaid the intine. The surface of the tectum carries many short spinules, and is either unpatterned or slightly sculptured, revealing a low reticulate pattern with phase contrast observation. The single aperture is operculate, the operculum itself having a stratification similar to that of the non-apertural exine. The earliest electron microscopic investigations were those of Rowley and collaborators (Rowley, Mühlethaler and Frey-Wyssling, 1959; Rowley, 1960, 1964). These showed that the two layers of the exine, the outer referred to as the ektexine and the inner the endexine, are each perforated by numerous channels, 14–25 nm in diameter. The surface spinules were shown to range in height between 0·07 and 0·2 μm, and in basal diameter between 0·08 and 0·3 μm. The spinules were found not to be perforated, and this was also true of the columellae (bacula) linking the tectum and the foot layer. Rowley (1960), in a pioneering study combining the use of surface replica and section techniques, compared the pollens of various wild grass species with those of certain cereals and cereal relatives, concluding that the ektexines of *Coix lachryma-jobi* L., *Zea mays* L. and *Secale cereale* L. had no surface patterning other than that given by the spinules, while the exines of *Poa* and *Phleum* species carried an incised reticulum, with one or more spinules per lacuna.

The first comprehensive developmental study of a grass exine was that of Skvarla and Larson (1966) on *Zea mays*. The fully mature exine was not covered in this study, but the features observed in electron micrographs of the maturing exine at the late-vacuolate microspore stage were in general agreement with those described by Rowley (1960, 1964) in *Zea* and other genera at maturity. Skvarla and Larson (1966) did, however, report an additional stratum, interposed between the foot layer and the intine, which they referred to as the endexine. More recent developmental studies have included those by Christensen, Horner and Lersten (1972), and Christensen and Horner (1974) on *Sorghum bicolor* (L.) Moench. These authors described a stratification of the non-apertural exine closely comparable with that of *Zea*, with a tectum subtended by bacula from the nexine, the latter overlying in turn a thin granular layer corresponding to the endexine of Skvarla and Larson (1966). Both the tectum and nexine were found to be perforated by channels, but these do not penetrate the granular layer. The terminologies applied by various authors to exine stratification are summarized in Fig. 1.

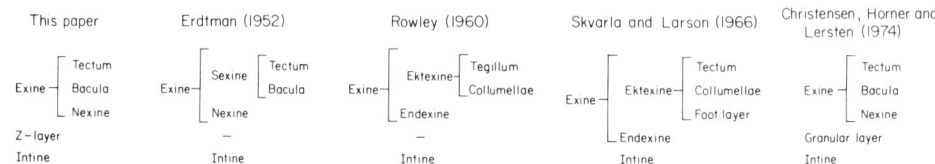

FIG. 1. The terminologies applied by various authors to the stratigraphy of the grass pollen wall.

The development of the germination aperture of the grass pollen grain has been traced in *Poa annua* by Rowley (1964), in *Zea mays* by Skvarla and Larson (1966), and in *Sorghum bicolor* by Christensen and Horner (1974). The essential features are similar in all three genera. The aperture is surrounded by a thickened annulus, the inner face of which is lamellate. The central operculum occupies about one-third of the area of the aperture, and is connected to the margin of the encircling annulus by a thin membrane (Plate 1A). This membrane overlies a thickened boss, which usually appears of intermediate electron density with normal fixation and post-staining procedures, occasionally with a fine granularity. This zone is continuous with the endexine of Skvarla and Larson (1966), and the thickening underlying the aperture is referred to as the 'Zwischenkörper' by Christensen and Horner (1974). This name will be retained in the present account, and to acknowledge the continuity, the endexine of Skvarla and Larson will be referred to as the Z-layer (Fig. 2).

The chemistry of the grass exine is yet to be investigated in any detail, but the cytochemical evidence suggests that the material of the sexine and nexine belongs to the general class of sporopollenins (Brooks and Shaw, 1971). The two layers cannot be differentiated by any optical staining procedure and show the same reactivity to electron stains. This implies a chemical homogeneity, and a lack of the differentiation observed in many other pollen exines (Faegri, 1956).

The chambered exines of many angiosperms are known to receive materials from the anther tapetum during the later stages of pollen development (Heslop-Harrison, 1968; Heslop-Harrison, Heslop-Harrison, Knox and Howlett, 1973: for reviews, see Heslop-Harrison, 1975, and Knox, Heslop-Harrison and Heslop-Harrison, 1975). The transferred material include proteins and lipids, the former usually inserted first, with the latter apposed to form an outer coat, often acting as a sealant for the micropores of the exine where this is tectate (Heslop-Harrison, 1979b). None of the electron microscopic studies of grass exines, however, gives any evidence of a massive transfer of tapetal proteins to the exine cavities, nor does the pollen bear a lipidic coating at maturity. The exine cavities usually appear wholly empty after the beginning of intine growth (e.g. de Vries and Ie, 1970), or at the most to contain vaguely defined strands of fibrillar material (e.g. Christensen and Horner, 1974).

In the mature grass pollen grain, the *intine* forms a continuous layer within the exine, often thickened somewhat at the apertural pole of the grain, and especially so in the vicinity of the aperture itself. Light microscopic cytochemistry has led to the conclusion that in the grasses, as in other angiosperms, the principal component of the intine is polysaccharide (Sitte, 1953). Electron microscopic images usually show the ground substance as homogeneous or vaguely microfibrillar, the latter appearance usually interpreted as indicating a cellulosic component (e.g. Christensen and Horner, 1974, Fig.

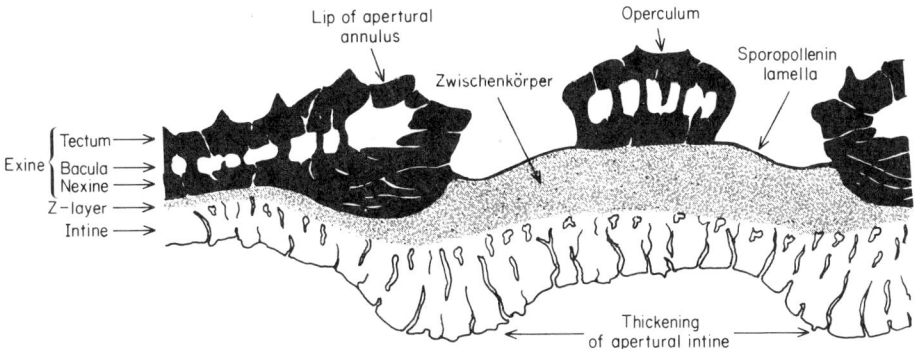

Fig. 2. Diagram of the aperture region of a grass pollen grain, identifying the various structural features referred to in the text.

49). The most conspicuous structural feature is a system of radially oriented channels, seen in the mature grain to extend from the neighbourhood of the plasmalemma of the vegetative cell towards the Z-layer. These channels were first observed by Rowley (1960), who realised that they arose during development from the peripheral cytoplasm of the vegetative cell. Developmental stages showing this origin have been illustrated by Christensen and Horner (1974) in *Sorghum*, and excellent aspects are to be seen in the electron micrographs of de Vries and Ie (1970) of developing pollen of *Triticum aestivum*.

In all of the angiosperms so far investigated, the intine of the pollen grain carries a protein component including several hydrolytic enzymes (review, Heslop-Harrison, 1975). The protein is inserted during the growth of the intine layer, the synthesis being in the peripheral cytoplasm of the vegetative cell (Knox and Heslop-Harrison, 1970; Knox, 1973). In monocotyledons in general the intine inclusions arise as microvillus-like extensions of the plasmalemma, which become invested in the polysaccharide of the intine and eventually cut off into the growing wall. The process is particularly well seen in the Iridaceae and related families, where the inclusions are often wide and very regularly spaced (Heslop-Harrison, 1975; Y. Heslop-Harrison, 1977). The microvilli are invested at first by the plasmalemma, but after severance from the cell surface the membrane is usually degraded, leaving the protein content free within a cavity of the polysaccharide wall. Electron-microscopic methods for the localization of the wall-held enzymes show that they are carried in these cavities (Knox and Heslop-Harrison, 1971a). Proteins are readily detectable in the intine of the grass pollen grain, and acid phosphatase, ribonuclease and non-specific esterase have been localized in this layer of the wall by optical microscopic cytochemical methods (Knox and Heslop-Harrison, 1970). The analogy with other monocotyledons, notably the Iridaceae, suggests that the enzymes are held in the intine cavities. As in all aperturate pollens, enzyme activity is greatest in the poral region, where it is associated with a thickened intine. Grass pollens emit a considerable spectrum of antigens on moistening, some of the antigenic activity being accounted for, no doubt, by the wall-held proteins. Localization by the immunofluorescence method shows that the antigens are indeed mostly derived from intine sites, the greatest concentration being, again, in the apertural intine (Knox and Heslop-Harrison, 1971b).

In the present work the structure of the wall of the pollen of *Secale cereale* (rye) has been investigated in detail, with the aim of establishing its functions during hydration and germination. Cytochemical and other methods have been used to clarify certain features of the chemistry of the wall, and to throw further light on the stratification, particularly in the region of the aperture. Special attention has been given to the fate of the Zwischenkörper and the apertural intine during germination and the formation of the tip of the emerging pollen tube.

The vegetative cell

The most conspicuous component in the vegetative cell cytoplasm is the stored starch, which at maturity is present in a quantity sufficient to mask most other features of the cell except the vegetative and sperm nuclei. For this reason, accounts based upon optical microscopy are, in the main, uninformative. A general review of the information available up to 1967 for maize and wheat was given by Goss (1968), and several pertinent Russian publications have been summarized by Poddubnaya-Arnoldi (1976) and Batygina (1974).

No comprehensive electron-microscopic studies of the protoplast of the mature grass pollen grain have been published hitherto. The accounts given by Skvarla and Larson (1966), de Vries and Ie (1970), Christensen, Horner and Lersten (1972) and Christensen and Horner (1974) of pollen development in various cereals are mainly concerned with the wall, and trace events up to the late vacuolate stage, after which the wall is essentially in its mature state; they do not, therefore, cast light on the state of the organelles and membranes in the cytoplasm of the mature and partly dehydrated grain at the time of anther dehiscence.

In this study the cytoplasm of the mature pollen grain of rye has been investigated electron-microscopically using various fixation and post-staining procedures, and the changes occurring during hydration and early germination have been followed. Complementary observations have been made using optical-microscopic cytochemical methods.

The present work has not been concerned with the nucleus of the vegetative cell, nor with the behaviour of the generative cells during germination.

Hydration and germination

In the grasses, fresh mature pollen grains germinate very rapidly on suitably receptive stigmas, a fact established by the early observations of Randolph (1936) on *Zea mays* and Pope (1937) on *Hordeum vulgare*. In general, germination in the small-grain cereals takes place in under 5 min; the more recent papers reporting this include, for oats (*Avena sativa* cvs), Brown and Shands (1957); for wheat (*Triticum aestivum* cvs), Watanabe (1955, 1961), Hoshikawa (1960), Kihara and Hori (1966), Zeeven and van Heemert (1970), Meyer (1971) and Lange and Wojcechowska (1976), and for barley (*Hordeum vulgare* cvs), Luxova (1967). The period for germination is also 5 min or less for pollinations of wheat by rye (Zeeven and van Heemert, 1970; Meyer, 1970; Lange and Wojcechowska, 1976). A still shorter period, within 1 min, is recorded for *Triticum aestivum* cv. 'Sonora' by Chandra and Bhatnagar (1974). On the other hand, somewhat longer periods have been given for rice (Cho, 1956), and the estimates given for maize are variable, ranging from comparatively short, 5–10 min (Randolph, 1936; Kihara and Hori, 1966), to very much longer (Goss, 1968).

The rapid transition of the male gametophyte from a state of comparative inactivity in the grain to one of vigorous growth indicated by many of the records for the germination of cereal pollens implies an astonishingly rapid re-establishment of a normal metabolism, including notably the synthetic capacity required for the synthesis of the pollen tube wall, a major activity of the gametophyte during its growth into the stigma. The activation of the grain depends on rehydration, and this is dependent on the inflow of water from the stigma after attachment of the grain. The speed with which the appropriate water balance is reached will naturally depend both on the capacity of the stigma to support the flow and the requirement of the grain. Estimates of the water content of grass pollens mostly refer to loss on oven drying to constant weight, and the values recorded vary widely. Early values for *Zea mays* range from 3·97 per cent (Vinson, 1927) to 65 per cent (Knowlton, 1921). More recent reports for the small grain cereals lie mostly in the range 15–40 per cent, but again with substantial variation even for the one species.

Two factors appear to contribute to the diversity of recorded values: (a) variation in the water content of grains at the time of release from the anther, and (b) differences in testing procedures, particularly in the period allowed for evaporative drying of the grains before measurements of water content are made. The first factor reflects developmental differences that may have some considerable significance for the function of the pollen. The second, procedural, factor contributes to the confusion in the literature, but it also reflects an important feature of the physiology of grass pollen, namely that it undergoes rapid and severe desiccation in the atmosphere after release from the anther.

Angiosperm pollens invariably undergo some dehydration in the anther during the final period of maturation (Linskens, 1967). In the grasses as exemplified by the small grain cereals, water is withdrawn from the anther loculi during the 24–48 h preceding anthesis, and the final desiccation takes place when the filament extends at anthesis. The pathways of water withdrawal from the individual grains are likely to be both through the intine at the germination aperture and the micropores of the sexine and nexine elsewhere (Fig. 2). At some point the continuity of water films in the anther loculi is broken, and further loss will then be evaporative, both through the aperture and the exine (Heslop-Harrison, 1979b). The rate of water extraction before this stage is reached will evidently depend on the steepness of the water potential gradient from the grains, into the anther wall, connective and filament, and thence into the plant. The general water status of the plant will therefore be a factor determining the extent of desiccation of the pollen at the time of release, and this has been observed in experiment. In a valuable recent paper, Barnabas and Rajki (1976) reported that the pollen of *Zea mays* shed in the morning in an ambient temperature of 18 °C and a relative humidity of 75 per cent contained as much as 45–60 per cent water, while that shed from plants of the same genotype at noon in a temperature of 24 °C and a relative humidity of 63 per cent contained 25–35 per cent water.

Barnabas and Rajki (1976) also gave values for the loss of weight of maize pollen in normal atmospheric conditions after release from the anther. Their data are expressed graphically in Fig. 3. Clearly most of the variation among reported values for the water content of maize pollen could be accounted for by difference in the holding time before determinations were made after anthesis. The same explanation might be sought for the wide differences in water content reported for other grass pollens.

In the experiments of Barnabas and Rajki (1976), the germinability of pollen on the stigma fell with progressive desiccation, reaching 9 per cent after 24 h, when the mean water content was 4·9 per cent (Fig. 4). Pfundt (1910) appears to have been the first author to appreciate the importance of atmospheric humidity in determining the longevity of pollens in storage, and the importance of this factor for the grasses was acknowledged by Knowlton (1921). Blair and Loomis (1941) observed that the germinability of maize pollen fell to zero after 18 h storage at 70 per cent relative humidity at room temperature, but remained at control levels after 30 h storage at 98 per cent relative humidity. The results of Sartoris (1942) were in general agreement; maize pollen stored at 4 °C in 90–100 per cent relative humidity retained the germinability of fresh pollen for 6–7 days. Sartoris also showed a clear relationship between relative humidity and germinability in pollen of sugar cane (*Saccharum officinarum* L.) stored at 4 °C; germinability was lost altogether in under 24 h in storage over fused calcium chloride, but was retained at control levels for 8 days in a saturated atmosphere. Similar results have been reported more recently for maize by Pfahler and Linskens (1972). Indeed, the evidence for maize and sugar cane

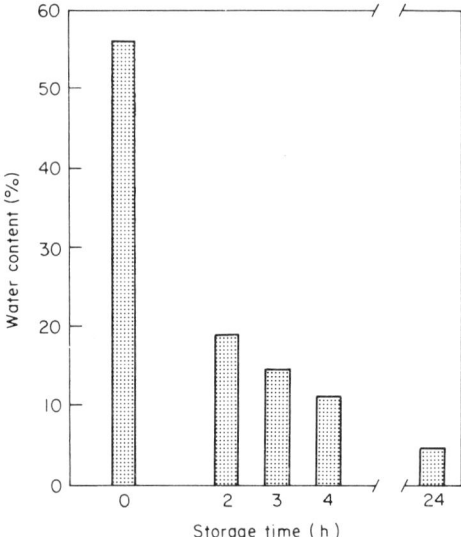

FIG. 3. Fall in water content of the pollen of maize during storage in normal atmospheric conditions. Data of Barnabas and Rajki (1976).

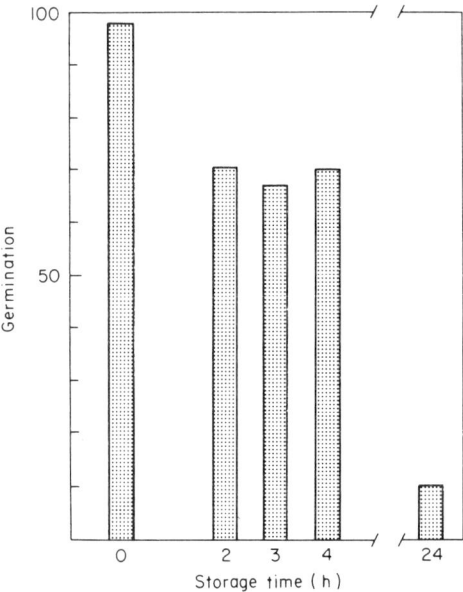

FIG. 4. Decline in germinability in maize pollen samples stored as in Fig. 3. Data of Barnabas and Rajki (1976).

seems unambiguous, and fragmentary results from other grasses also support the conclusion that desiccation of the pollen leads to a rapid loss of viability. However, a striking exception has been reported by Vasil (1960), who found that the pollen of *Pennisetum typhoideum* (L.) Rich. stored at 16–35 °C over concentrated sulphuric acid showed 78 per cent germination after 186 days.

Results such as those of Barnabas and Rajki (1976) suggest that under normal atmospheric conditions the highest germinability in grass pollens is likely to be found immediately after shedding, and this is indeed the general experience. However, several authors have reported that the percentage germinability may increase after a period of storage in a suitable environment, usually in an atmosphere approaching saturation (e.g. Blair and Loomis, 1941; Pfahler and Linskens, 1972). This may indicate that under favourable conditions ripening continues after release, possibly by the continued conversion of storage starch into sugar. The likely importance of the starch/sugar balance has been noted by various authors (e.g. Niethammer, 1932). Watanabe (1961) found that pollen extracted from anthers of wheat and other grasses a short period before natural release was incapable of germinating on a receptive stigma. Such pollen usually shows a uniform distribution of starch in the vegetative cell, and Watanabe observed that as the grains become germinable a starch-free zone appears towards the apertural pole. This may indicate a fall in starch content and a concomitant increase in sugar, but as yet there have been no direct measurements of the sugar content of germinable and non-germinable grains to support the contention.

It is obvious enough that for mature, germinable pollen the period from capture to the emergence of the tube will be affected by the degree of hydration of the grains at the moment of receipt. The expected relationship is simply that the more fully hydrated the grain, the shorter the interval before germination. Within the limits set by the progressive loss of viability, desiccation should protract the interval before tube emergence, the additional time representing the longer period needed for water uptake. Results reported in the literature suggest that this is true. However, with viable pollen the hydration on the stigma surface is not accompanied by a continuous dilation of the grain, for the initial period of enlargement is followed by an interval when liquid is exuded both from the aperture and the non-apertural exine. This exudation was first noted by Watanabe (1955),

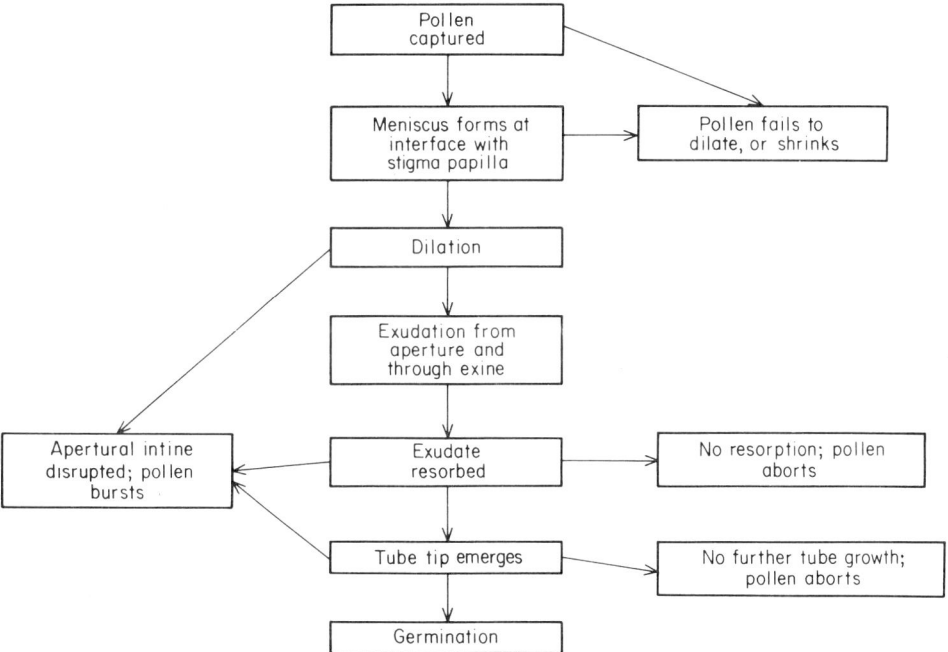

FIG. 5. Scheme setting out the possible events following pollination the grasses, based upon a diagram from Watanabe (1961). The spinal sequence shows the episodes in a normal successful pollination. The lateral arrows indicate points where the sequence may be broken and the pollination aborted.

who observed it in 28 species of 19 genera of the Gramineae. Watanabe (1955, 1961) established that the exudation is an essential prelude to germination in the grasses; grains that do not exude, do not germinate – although it does not necessarily follow that all grains that do exude will eventually germinate. Figure 5, re-cast from a diagrammatic scheme given by Watanabe (1955), sets out the normal sequence in a successful pollination, and shows how the process may be aborted at any of the successive stages by premature desiccation, or by the bursting of the grain through the disruption of the apertural intine before the emergence of the tube tip.

In a later section (p. 34) an interpretation of the hydrodynamics of germinating pollen is given; this is based upon certain conclusions concerning the state of the membranes of the vegetative cell of the male gametophyte at the time of dispersal and the transformations they undergo after contact with the stigma (Heslop-Harrison, 1979a). It is likely that the membranes of the partly dehydrated grain are not effective as barriers to the movement of solutes, and do not become so until hydration has advanced to a critical threshold. It may be seen, then, that both the first period of pollen hydration and the succeeding exudation phase are critically important in the early interactions with the stigma. The initial hydration, beginning with the formation of the meniscus between the grain and the stigma surface, is likely to be the interval in which the stigma secretions pass into the grain. During the exudation phase, the net movement is outward from the grain surface not immediately in contact with the stigma, and it is probably during this interval that much of the emission of mobile protein and other surface- or wall-held materials takes place (Knox and Heslop-Harrison, 1971b; Heslop-Harrison, Heslop-Harrison and Barber, 1975).

The present paper includes a detailed account of the sequence of events accompanying hydration, exudation and the emergence of the tube during the germination of rye pollen.

MATERIALS AND METHODS

Plant material

The main observations were made on *Secale cereale* L. cv Rheidol (rye). The plants were grown out of doors in the Royal Botanic Gardens, Kew, and at the Welsh Plant Breeding Station. Supplementary observations were made on *Gaudinia fragilis* L. (Beauv.), *Poa trivialis* L., *Dactylis glomerata* L. and *Hordeum bulbosum* L.

Influorescences in early anthesis were brought into the laboratory for the collection of pollen, and to promote anthesis they were exposed to incandescent lamps or to a Philips 300 W infra-red source.

Pollen quality was estimated using the fluorochromatic reaction (Heslop-Harrison and Heslop-Harrison, 1970). The stock substrate solution contained 2 mg ml^{-1} fluorescein diacetate in acetone, and this was added dropwise to the suspension medium containing 10–20 per cent sucrose until milkiness appeared. The proportion of grains showing fluorochromasia was recorded after 5–10 min.

The timing of pollen hydration and germination on the stigma was carried out with single-grain pollinations. Virgin receptive stigmas were detached from opening flowers and mounted in pollination cells made from glass slips on a microscope slide. The base of the stigma was wetted with a filter-paper wick, but the stigma itself was unenclosed to prevent condensation. The pollen grains were placed individually in the desired orientation on selected stigma papillae with a human eyelash hair. The preparations were observed using epi-illumination, or combinations of epi- and transmitted illumination, with Vickers Photoplan or M17 systems. Micrographs were taken on Ilford Pan F film at the appropriate intervals, timed manually, or by an electronic timer controlling an automatic camera.

Grass pollens are not readily germinated *in vitro*, and despite extensive trials no wholly satisfactory method has been found in the present work for obtaining consistent germination of rye pollen in liquid media. Ahloowalla (1973) has reported good germination of pollen of *Lolium perenne* in a liquid medium containing sucrose, calcium chloride and boric acid, but this medium did not support the germination of rye pollen. Similarly, germination was not obtained in media containing polyethylene glycol (Ferrari and Wallace, 1975) as osmoticum.

Consistently good germination was obtained on a semi-solid medium. This contains the minimal constituents now known to be required for angiosperms in general, namely a sugar, calcium and boron (Vasil, 1964). The optimal medium for rye (Medium A) contains 1 per cent agar or agarose (BDH), 0·6 M sucrose, 10^{-3} M boric acid and $1-2·5 \times 10^{-3}$ M calcium chloride or nitrate (Shivanna, Heslop-Harrison and Heslop-Harrison, 1978). Germination begins in 60–70 s on this medium.

Replacement of the sucrose in Medium A with another osmoticum – for example, mannitol or pentaerythritol (Dickinson, 1967) – inhibited germination; and other sugars such as lactose and glucose did not substitute adequately for sucrose. The boron concentration of the medium is not critical, but the calcium concentration was found to affect both germination rate and tube growth (Fig. 11). Supra-optimal calcium concentrations prevent the emergence of the pollen tube while permitting normal hydration and activation of the grain. The experiments on this effect were carried out with a medium containing 1 per cent agarose, 0·6 M sucrose, 10^{-3} M boric acid and 10^{-2} M calcium nitrate (Medium B).

The hydration of the surface of semi-solid medium is critically important in obtaining consistent germination and tube growth (Bar-Shalom and Mattsson, 1977). Plates of Medium A taken directly from storage have a wet surface, and on this germination is poor, with many grains bursting. Germination is improved when a firm film is allowed to form over the surface by drying out for 10 min or so in laboratory air as recommended by Bar-Shalom and Mattsson (1977), but it is difficult to ensure consistent drying by this method. In the method finally adopted, the medium was poured to a depth of 1–1·5 mm in non-wettable polystyrene petri dishes. After cooling, the gel was cut into portions up to 10 mm square, and immediately before use the segments were lifted, using a glass coverslip to avoid metal contamination, and placed *inverted* on a microscope slide. The lower surface – that formerly in contact with the base of the petri dish – was then used as the germination surface. The hydration of the lower face is less variable than the upper, so the method gives much more consistent results.

Darlington and La Cour (1960) describe the use of floating cellophane films on appropriate medium as a surface for the germination of certain pollens. The method does not give good germination of grass pollens, but it was used to give controlled hydration of grains for following exudation phases and starch conversion. Squares of single-layer Visking tubing were floated on Medium A, and the pollen sown on to the surface; the preparations were kept uncovered during observation to prevent condensation.

Uncovered preparations were used for microscopy with objectives up to $20 \times$. For higher magnifications, samples were flooded with liquid paraffin (BDH) and a coverslip applied. The tubes survive and grow freely in these mounts for up to 3 h.

Preparation of pollen eluates

Eluates for the estimation of protein release from hydrating pollen were prepared from samples of 200–500 mg. These were dispersed in 2–10 ml 0·4 M sucrose, and after appropriate periods the grains were separated by centrifugation at 7000 g for 3 min. Second and subsequent eluates were made by re-dispersing the pollen in the same volume of medium and centrifuging once more. The period required for dispersal is about 30 s, and this set the minimum time interval for following the outflow from the samples.

Some small further elution would presumably take place during centrifugation. After the completion of the elution, 40–60 per cent of the grains showed fluorochromasia (Heslop-Harrison and Heslop-Harrison, 1970), suggesting that 50–75 per cent of the viable grains of the original samples survived the treatment with intact membranes.

Eluates for the estimation of carbohydrate released were prepared in the same manner, using 0·25 M $CaCl_2$ as the medium.

Protein content of eluates

The protein content of eluates was estimated by the Lowry procedure, with bovine serum albumen as a standard. Acid phosphatase activity was measured with *p*-dinitrophenol phosphate as substrate. The reaction mixture contained 0·25 mg substrate in 1·5 ml 0·1 M citrate buffer at pH 5·0 with 0·1 ml eluate. Incubation was for 30 min, and extinction was measured at 410 nm after the reaction had been stopped by the addition of 4 ml 0·1 N NaOH.

Proteins of pollen and stigma eluates were labelled with fluorescein isothiocyanate (FITC) as described by Vithanage and Heslop-Harrison (1979).

Carbohydrate content of eluates

Total carbohydrate in the eluates was estimated with naphthol-H_2SO_4 reagent, with sucrose as a standard. The sugars present were separated by thin-layer chromatography on silca gel plates impregnated with 0·2 M Na_2HPO_4 and activated for 1 h at 110 °C. Satisfactory separations of arabinose, xylose, galactose, sucrose and glucose was given by solvent No. 37 of Lato, Brunelli and Giuffini (1969), acetone:water:chloroform: methyl alcohol (16:1:2:2, by volume). The sugars were detected by spraying with (*a*) naphthoresorcinol reagent (20 mg naphthoresorcinol in 10 ml ethanol containing 0·3 ml conc. H_2SO_4) or (*b*) 3 per cent *p*-anisidine, followed by heating at 100 °C. Tentative identifications were made by R_F and colour reactions.

Dissolution of pollen exines and preparation of intine 'ghosts'

Bailey (1960) reported that the sporopollenin of the pollen exine is soluble in 2-ethanolamine, and Southworth (1974) has investigated the effects of this solvent on the exines of several angiosperms, including that of grass pollens. In all of the species tested, Southworth found that the exine underwent some dissolution on heating in 2-ethanolamine, although in some instances the nexine was not completely dispersed. In the present work, the exines of all the grasses tested were found to be soluble in both 2-ethanolamine and 3-ethanolamine, the dissolution proceeding rapidly as the temperature was raised to 100 °C. This treatment left distorted and disrupted grains, however, and dissolution with less damage was obtained by solvation over a period of 12–36 h in 2-ethanolamine at 80 °C. No cytochemically detectable sporopollenin remained after this extraction.

After the removal of the exines, the grains were recovered by diluting the solvent with 10 × the volume of distilled water and centrifuging. The residual solvent was removed by resuspending in water and centrifuging three further times.

Exhaustive extraction with 2-ethanolamine does not destroy the structural integrity of the intine, but is likely to remove both hemicellulose and pectic components (Bouveng, 1965). The intine was not further affected when the washed, exine-less grains were digested in N NaOH at 60 °C for 12 h. This treatment eliminated the protoplast, leaving intine 'ghosts'. Alkaline digestion might be expected to remove any pectic materials surviving the 2-ethanolamine extraction.

Intines were also recovered from the exine-less grains mechanically, by shearing a suspension between glass plates until inspection showed that a high proportion of the intines had been rolled free of the protoplasts.

Intact intines cannot readily be obtained from *whole* grains, but fragments can be separated from the exine by mechanical shearing. These fragments proved useful in providing a comparison with the chemically isolated intines in staining and other properties.

Chemically isolated intines were subjected to digestion by cellulase (two sources, 'Cellulase 500', Onozuka; Cellulase, from *Aspergillus niger*, Sigma), pectinase (from *A. niger*, Sigma), and Helicase. The conditions are given in the footnotes to Table 3.

Preparation for optical microscopy

For enzyme localization, fresh pollen was encased in 15 per cent gelatine and sectioned at 4–8 μm on a Slee cryostat (Knox and Heslop-Harrison, 1970). Some observations were also made on material fixed in 3 per cent glutaraldehyde buffered in 0·1 M phosphate buffer at pH 7·2, washed, dehydrated through an alchohol series and embedded in low-melting-point wax (Ralwax, Raymond Lamb Co.) at 42 °C. This was found to section satisfactorily down to 4 μm in the cryostat at 2–6 °C.

For high-resolution optical microscopy, samples were fixed in 1·5 or 3 per cent glutaraldehyde in 0·05 M phosphate buffer at pH 7·2, washed, dehydrated through an alcohol series and embedded in JB4 resin (Polysciences Inc.). Sections were cut in the thickness range 0·5–1·0 μm with glass knives on a Porter-Blum MT-1 ultramicrotome and mounted directly on glass slides. Sections prepared in this medium were found to stain readily with all standard procedures using aqueous media, including those for fluorescence microscopy.

Enzyme localization

Non-specific *esterase* was localized in freeze-sectioned material using α-naphthyl acetate (Sigma) as a substrate in coupling reactions with Fast Blue B or BB salts (Pearse, 1972). *Acid phosphatase* was localized by a variety of methods (Knox and Heslop-Harrison, 1970). The highest resolution in wall sites was given with α-naphthyl acid phosphate (Sigma) and hexazotized pararosanilin as a coupling agent (Barka and Anderson, 1962). Controls were run without substrate, or after pre-incubation with 0·01 per cent NaF. *Ribonuclease* was localized as described by Knox and Heslop-Harrison (1970).

Pectinase activity was detected in pollen and pollen-tube exudates by a substrate film method. Germination Medium A was modified by reducing the agarose content to 0·75 per cent and incorporating 0·25 per cent apple pectin (BDH). Thin films of this medium were cast on microscope slides, and pollen samples were incubated on these in a moist chamber at 25 °C for periods of 10 min to 1 h. The pectin in the films was detected by immersing the slides in warm 1 per cent hexadecyltrimethylammonium bromide (Sigma) for 20–30 min and viewing the precipitate produced in the gels with dark-ground illumination. Pectinase activity around grains and tubes produced zones free from the precipitate.

Protein localization

Conventional methods were used for the staining of proteins in sections, including naphthol yellow S (Deitch, 1955), mercuric bromophenol blue (Mazia, Brewer and Alfert, 1953), amido black (1 per cent in 7 per cent acetic acid), and Coomassie blue (water:methyl alcohol:acetic acid, 87:10:3 by volume, with 0·25 g Coomassie Brilliant Blue G per 100 ml).

The same Coomassie blue stain-fixing medium was used for the detection of proteins in pollen exudates and in smears (Heslop-Harrison, Heslop-Harrison, Knox and Howlett, 1973). Some observations were also made using the fluorescent protein stain 1-anilinonaphthylsulphonic acid (1-ANS) (Heslop-Harrison, Knox and Heslop-Harrison, 1974). The staining medium contained 1-ANS at 0·001 per cent in 0·01 M phosphate buffer at pH 6·8 with 15 per cent methanol.

Lipid localization

Lipids in sections were detected by standard methods using Sudan III and Sudan black (Jensen, 1962). More satisfactory localization in 0·5–1 μm sections of JB4 embedded material was obtained with scarlet R, used as specified by Gurr (1965).

Pollen wall cytochemistry

Sporopollenin, the characteristic material of pollen exines, has been partly characterized as a polymer or co-polymer of carotenoids and carotenoid esters (Brooks and Shaw, 1971). No specific cytochemical staining procedures are available for sporopollenin, but it is usually accepted that those parts of spore and pollen grain walls that are resistant to acetolysis belong to the general class of sporopollenins. For isolating the exines of rye pollen, samples were acetolysed by warming in acetic anhydride:concentrated sulphuric acid (9:1, by volume) until the cell contents were fully removed (Erdtman, 1960). The exine was detected in sections and unacetolysed preparations by staining in auramine O, 0·01 per cent in 0·05 M Tris–HCl buffer, pH 7·2 (Heslop-Harrison, 1977) and observed by fluorescence microscopy. With this staining, the sporopollenin of the grass exine fluoresces a brilliant orange-yellow; the stain also has affinity for cutin and lipids, but these have a greenish yellow fluorescence when observed with Vickers exciter filter No. 1 and barrier filter No. 3.

Since the work of Sitte (1953) it has been known that pollen and spore intines have a microfibrillar cellulose component, and chemical evidence for the presence of cellulose in the intines of vascular plants has been given by Shaw and Yeadon (1966). Hemicelluloses and pectic substances are also present in the intine (e.g. Bouveng, 1965; Roland, 1971), which therefore has a composition broadly similar to that of the primary walls of somatic tissues (Northcote, 1975). The cytochemical methods currently available are inadequate for the unambiguous identification of the different cell wall polysaccharides *in situ*; and indeed simple chemical categorization is not itself especially useful in view of the structural complexity of the wall. However, an effort has been made to establish some of the characteristics of the intine of rye pollen using a combination of staining and extraction procedures. The periodic acid–Schiff (PAS) procedure was used for the general localization of polysaccharides in sections and isolated cell components (Pearse, 1972). As a sensitive and effective fluorescent stain for wall polysaccharide, calcofluor white MR2 (Polysciences Inc.) was used at a concentration of 0·001 per cent in aqueous solution. Idle (1966) appears to have been the first user of fluorescent whiteners for the microscopy of higher plant tissues, and calcofluor white has been extensively employed subsequently, notably for the early detecting of wall material forming around isolated protoplasts (e.g. Nagata and Takebe, 1970). Chemical evidence suggests that the stain has affinity with 1,3- and 1,4-linked glucans (Maeda and Ishida, 1967; Takeuchi and Komamine, 1978), in which case it might be expected to reveal cellulose. In the present work calcofluor white was found to stain intine ghosts from which hemicelluloses and pectic materials had been removed by exhaustive chemical extraction; such ghosts showed birefringence, and presumably represented the residual cellulosic skeleton of this layer of the wall.

Pectic substances were localized by alcian blue and ruthenium red staining. The mechanism of alcian blue staining has been discussed by Lev and Spicer (1964) and Pearse (1972). At pH 2·0–2·5 the dye reacts mainly with uronic acid groups; in plant cell walls it is usually regarded as staining pectic acids, including gel-forming components of certain surface and intercellular secretions. The dye (Alcian Blue 8GX, R. A. Lamb) was made up at 1 per cent in 3 per cent acetic acid, giving a solution with a final pH of 2·5. Ruthenium red has been used as a stain for cell surface polysaccharides by Luft (1964), and for acid 'mucopolysaccharides' by Pihl, Gustafson and Falkmer (1968). It was used in 0·02 per cent aqueous solution following Gurr (1965).

Callose was identified by its fluorescence in decolorized aniline blue. This staining unambiguously identifies callose in the original sense of Mangin (1889) in pollen tetrads and sieve tubes (Eschrich, 1961; Eschrich and Currier, 1964). Doubts have been expressed as to whether the wall compound so revealed is necessarily a homogeneous β-1,3-glucan, which, since the work of Kessler (1958), has tended to become the chemical definition of 'callose'. Callose in this sense may be something of a conceptual artefact, since most callose preparations from plant cell walls have components with 1,3- and 1,4-linkages. However, in pollen and pollen tubes, the wall layers showing fluorescence after staining in decolorized aniline blue are easily distinguishable in the electron microscope from microfibrillar cellulosic wall layers by their homogeneity (Plate 13 B); they are therefore distinct structural components. The presence of the callosic inner layer in the pollen tube provides the basis for localizing tubes in stylar tissues by their fluorescence after aniline blue staining, a technique introduced by Linskens and Esser (1957). For callose localization in rye pollen, the dye (Merck or BDH) was used at 0·05 per cent at pH 11.

Electron microscopy

Material for transmission electron microscopy was fixed 1–4 h in 1·5 or 3 per cent glutaraldehyde in 0·05 M phosphate buffer at pH 7·2 containing 8–12 per cent sucrose. The sucrose concentration required to prevent bursting of pollen grains or tubes was established by observing the effects with the optical microscope. Material was post-fixed in 1 per cent OsO_4 in the same buffer for 1·5–2 h at c. 4 °C. The fixed material was washed in buffer and distilled water, dehydrated through an alcohol series, transferred to propylene oxide, and then embedded in Araldite by standard procedures. Thin sections were cut with diamond knives on a Huxley Mk III ultramicrotome.

The standard post-staining procedures with uranyl acetate and lead citrate provided satisfactory images of the pollen and pollen-tube walls, but were quite inadequate for revealing many cytoplasmic details in ungerminated grains or during the early emergence of the tube. In particular, cytoplasmic and organellar membranes were poorly defined, even after the full hydration of the pollen grain. Among the variety of staining procedures tested, the use of potassium permanganate, alone, or following uranyl acetate staining, was found to be the most effective. Grids were immersed in distilled water in plastic capsules with a perforation in the bottom, and the capsules and contents transferred quickly into freshly made up 1 per cent potassium permanganate. The stain was allowed to diffuse into the capsules for periods of up to 15 min, and the capsules were then withdrawn and transferred without draining into a vial of distilled water and the grid decanted. By this means the need to pass the grids through a potassium permanganate solution/air interface was avoided.

Procedures using aqueous fixatives were virtually useless in stabilizing the emissions of hydrating and germinating pollen, which were dispersed quickly in contact with the fixing fluid. Partial stabilization was, however, achieved using osmium tetroxide vapour as a fixation. Crystals of osmium tetroxide were placed on cotton wool in the base of a small vial, and stigmas bearing hydrating or germinating grains were suspended at a height of a few millimetres above. The vial was sealed and left at room temperature until the tissue blackened, in some cases overnight. After fixation the stigmas were taken straight through the dehydration and embedding procedure without washing. Observations were made with an AEI Corinth electron microscope.

For scanning electron microscopy of intact grains, samples were dried by the critical point method and coated with gold–palladium. Observations were made with a JEOL microscope.

OBSERVATIONS AND EXPERIMENTAL RESULTS

The pollen wall

The exine

The surface of the exine and the germination aperture is seen in the scanning electron micrograph of Plate 1 A, of *Poa trivialis*. The stratification as observed by light microscopy is shown in Plates 3 A, B and 4 A, B, and by electron microscopy in Plates 9 A, 15 A and 16 A, B, C, all of rye.

TABLE 1. *Dimensions of exine features, derived from electron micrographs of tangential and radial sections of pollen fixed in OsO_4 vapour: Secale cereale cv. Rheidol*

	Thickness (μm)	Micropores		
		Diameter (nm)	No. per μm^2	Per cent area
Tectum	0·15–0·30	37·8 ± 0·45	54·19 ± 0·75	6·07
Nexine	0·12–0·25	38·1 ± 0·36	36·63 ± 0·84*	4·19
		Bacula		
		Diameter (nm)	No. per μm^2	
Lumen	0·15–0·25	119·3 ± 2·98	9·53 ± 0·15	

* Difference between tectum and nexine significant ($P < 0.005$). The spinules of the tectum are 0·12–0·18 μm in height, with a base width of 0·20–0·25 μm.

The dimensions of the cardinal features of the relaxed exine of rye are given in Table 1. The tectum, including spinules, is slightly thicker than the nexine. The dimensions of the bacula vary considerably, and no attempt has been made to calculate a mean diameter. Leaving the bacula out of account, however, it may be estimated that the lumen accounts for some 30 per cent of the total volume of the exine.

The staining properties of tectum, bacula and nexine are identical (Table 2), and also the electron opacities following standard staining procedures. This suggests that the sporopollenin is undifferentiated chemically throughout the exine, but the slight difference in solubility in 2-ethanolamine mentioned in a later section may indicate a different packing density in tectum and nexine.

The outer surface of the exine is bounded by a darker layer, c. 7 nm in thickness, separated from the homogeneous sporopollenin by an electron-transparent space of about the same dimensions (Plate 3 D). This characteristic configuration is present over the entire surface, including the spinules. A similar bounding layer is illustrated in the micrographs of Skvarla and Larson (1966) of *Zea*, and is commented on in *Sorghum* by Christensen *et al.* (1972). Nothing can be said about its nature, but since it is present in exines still in the process of thickening, it may simply reflect some characteristic mode of deposition of sporopollenin precursors.

TABLE 2. *Staining properties of the wall and certain components of the protoplast of ungerminated* Secale cereale *pollen*

| Stain | Wall | | | | Protoplast | | |
| | Exine | Zwischenkörper | Intine | | Starch | P-particles and cytoplasm[1] | Globuli[2] |
			In situ	Extracted			
PAS	0	0	++	(+)[3]	+++	+	0
Calcofluor white	0	(+)[4]	+++	+++	+++	++	0
Alcian blue	0	+++	+	0	0	(+)	0
Ruthenium red	0	++	+	0	0	0	0
Decolorized aniline blue	0	0	(+)[5]	(+)[5]	0	0	0
Toluidine blue	+++ (blue)	+++ (red)	0	0	0	(++) (purple/red)	?[6]
Naphthol yellow S	0	0	(+)[7]	0	0	(+)	0
Coomassie blue	0	0	(+)[7]	0	0	(++)	0
Amido black	0	0	0	0	0	(+)	0
Bromophenol blue	0	0	0	0	0	(++)	0
1-ANS[8]	0	0	0	0	0	(++)	(+)[9]
Auramine O	+++	0	0	0	0	(+)	(+)
Sudan IV	+	0	0	0	0	(+)	++
Sudan black	+	0	0	0	0	(+)	++
Scarlet R	+	0	0	0	0	(+)	++
Osmium tetroxide	++	0	0	0	0	(+)	++
Azure B	0	0	0	0	0	(++)	0

0, No appreciable staining; +, light staining; ++, moderate staining; +++, heavy staining.
[1] Unbracketed symbols, P-particle content; bracketed symbols, associated cytoplasm.
[2] At the surface of the vegetative cell and occasionally throughout.
[3] Aperture region only.
[4] Ambiguous because of the proximity of the intine.
[5] Scattered particles associated with the intine in occasional grains.
[6] Ambiguous because of the staining of the cytoplasm.
[7] Resolved only at the aperture in freeze-sectioned grains (Knox and Heslop-Harrison, 1970).
[8] 1-Anilinonaphthylsulphonic acid.
[9] Distinguished by bluish fluorescence.

Z-layer and Zwischenkörper

The Z-layer (= 'endexine'; see Fig. 1) in *Zea* has a thickness of about 40 nm (Skvarla and Larson, 1966). The corresponding stratum in rye is only visible in the vicinity of the apertures, but oblique sections of the exine–intine interface in the non-apertural wall show the characteristic – and not readily interpretable – appearance seen in Plate 3c. The greater electron opacity of the nexine near the interface might be accounted for were the thin Z-layer to be of higher electron-opacity, but this would not explain why the contiguous intine should be of *lower* opacity than the intine remote from the interface. The possibility that the appearance is an artefact of fixation or post-staining remains open.

In the vicinity of the apertures, the Z-layer is continuous with a thickened plug overlying the intine at the aperture, the Zwischenkörper of Rowley (1964) and Christensen and Horner (1974). An electron micrograph clearly showing this from the work of the latter two authors is reproduced in Plate 1b. In mature pollen fixed without preliminary hydration, the Zwischenkörper is seen in turn to be overlaid with a thin stratum of sporopollenin, continuous with the exine and operculum, and distinguishable by stainability with auramine O (Plate 4 B; Fig. 2). This layer is disrupted during the early hydration of the grain, and so does not always appear in material fixed in media of low tonicity. The Zwischenkörper itself has no affinity for the fluorescent stains auramine O and primulin, and lacks other of the cytochemical properties of sporopollenin (Table 2). It is dispersed by mild acetolysis, leaving the operculum suspended on the thin surface layer, if it survives *in situ* at all. The material of the Zwischenkörper is not therefore sporopollenin, and it is not to be regarded, developmentally, as part of the exine.

The staining reactions listed in Table 2 were observed in pollen walls separated mechanically from the protoplast in 20 per cent sucrose, as well as in embedded and sectioned samples. The results from the PAS reaction can probably be discounted, since periodic acid oxidation disperses much of the material of the Zwischenkörper. With calcofluor white, interpretation is difficult because of the intense fluorescence of the adjoining intine. However, the staining in alcian blue at pH 2·5 is not in doubt. The Zwischenkörper is clearly demarcated, both from the underlying intine and the overlying operculum (Plate 4 C, E, F). Stainability is lost after short extraction with ammonium oxalate. Similar, but much less intense, staining is given with ruthenium red. These reactions, taken together with the swelling properties of the Zwischenkörper described in a later section, suggest that it is a zone of concentrated, gel-forming pectin polysaccharides, dehydrated and so firmly sealing the aperture at the time of pollen dispersal (Heslop-Harrison and Heslop-Harrison, 1979).

Exine solubility

The dissolution of the exine in warm 2-ethanolamine is illustrated in Plate 5 A. The tectum first enlarges disproportionately, so tearing the bacula, and is then wholly dispersed. The nexine separates from the intine, and ultimately shares the fate of the tectum; the thickened apertural annulus is the last component of the exine to be removed.

The intine: structural features

The intine forms an irregular layer some 0·2–0·4 μm in thickness in the relaxed state in the rye pollen grain (Plates 4 A, 7 A, B, 9 A, B). The thickness is greater in the vicinity of the aperture, 0·4–0·6 μm. The profile at the aperture shows a thickening under the sporopollenin annulus surrounding the shaft and a prominence under the Zwischenkörper and operculum. Intine 'ghosts' show that this apertural differentiation is not imposed simply by shape of the overlying exine (Plates 5 B and 11 B). Surface views of the apertural zone reveal the greater thickness, and also show that the aperture is the focus of radiating lines of thickening extending over most of the surface area of the grain (Plate 11 A).

Viewed in profile, the apertural zone shows birefringence (Plate 5 C).

The intine: cytochemistry

The staining properties of the intine *in situ* and after chemical isolation are set out in Table 2, and the effects of various chemical and enzymic digestions in Table 3. *In situ*, the intine shows moderately strong PAS-reactivity, but intine ghosts are very much less reactive, and indeed staining is seen only at the thickened apertural site. In contrast, both intine ghosts and intines *in situ* show strong affinity for calcofluor white (Plates 4 A and 5 B).

Intines *in situ* show slight stainability with both alcian blue and ruthenium red, but this is lost from intine ghosts. In ungerminated grains, callose, as detected by fluorescence following decolorized–aniline blue staining, is very rarely present. When it does appear it is as scattered granules on the inner surface, and these may be seen also in intine ghosts, indicating that callose survives the isolation procedure.

TABLE 3. *Responses of the wall and certain constituents of the protoplast of ungerminated rye pollen to chemical and enzymic digestion*

Treatment	Wall			Protoplast	
	Exine	Intine		Starch	P-particles
		In situ	Extracted		
Chemical					
Acetolysis	R	D	D	D	D
2-Ethanolamine	D	R[1]	R[1]	R	R
Enzymic					
Cellulase[2]	R	D	D	R	?
Cellulase[3]	R	D	D	R	?
Pectinase[4]	R	R	D[5]	R	?
Amylase[6]	R	R	R	D	R
Helicase[7]	R	D	D	R	?

R, Resistant; D, dispersed by treatment; ?, result ambiguous, or could not be judged.
[1] Stainability in pectin stains reduced or lost.
[2] *Aspergillus niger* source, Sigma; 2 mg ml^{-1} at pH 5·6; incubation at 37 °C 6–12 h.
[3] Cellulase 500, Onozuka; 2 mg ml^{-1} at pH 5·6; incubation at 37 °C 6–12 h.
[4] *Aspergillus niger* source, Sigma; 2 mg ml^{-1} at pH 5·6; incubation up to 48 h at 37 °C.
[5] Loosened and slowly dispersed after prolonged incubation.
[6] Bacterial source; 10 mg ml^{-1} at pH 6·5; incubation 12 h at 25 °C.
[7] Two mg ml^{-1} at pH 5·6; incubation at 37 °C 6–12 h.

Cellulases from both of the sources tested completely destroyed intine ghosts, as did Helicase (Plate 5D, E). The pectinase from *Aspergillus niger* caused the intines to enlarge in short-term digestion, and to become partly dispersed after 48 h incubation.

The intine: protein inclusions

The intine tubules which are a conspicuous feature in developing grass pollens (seen, for example, in the micrographs of *Triticum* pollen of de Vries and Ie (1970) and of *Sorghum* pollen, Christensen and Horner, 1974) are less obvious in electron micrographs of ungerminated, mature grains, probably because of the desiccated state of the wall (Plate 7A). After full hydration and during germination they become more conspicuous. Profiles seen in tangential section frequently show what appear to be fragments of bounding membranes, presumably residues of the plasmalemma (Plate 3C). The tubules in rye are too irregular in distribution and diameter to attempt any assessment of total volume.

Knox and Heslop-Harrison (1970) used staining and cytochemical methods to locate proteins, including a number of acid hydrolases, in the intines of grasses. The methods adopted for enzyme location for optical microscopy produced a concentration of reaction product centred principally in the intine at the germination apertures, but the resolution obtained was inadequate to exclude the possibility that some activity might be present in the Zwischenkörper. High-resolution localization in freeze-sectioned pollen of rye shows that the main esterase and phosphatase activity is indeed centred in the intine (Plate 4D). The distributions seen in the micrographs of Knox and Heslop-Harrison (1970) of *Alopecurus*, *Coix* and *Lolium*, in which enzyme activity appears to extend right into and through the shaft of the aperture, are probably to be accounted for by diffusion of the wall-held enzymes into the Zwischenkörper region as the grain hydrated during preparation. The method for RNase, based on that of Enwright, Frye and Atwal (1965), is probably inadequate to give the resolution necessary to show which layer of the wall carries this enzyme.

Cytoplasm and organelles in ungerminated pollen

Amyloplasts

The starch in the fully mature but ungerminated pollen of rye is massed mostly in the hemisphere of the vegetative cell opposite to the aperture, leaving a starch-free zone adjacent to the aperture itself. This distribution is characteristic of germinable grains. As shown by Watanabe (1955), the starch is more generally distributed in grains at earlier developmental stages when the pollen is incapable of germination.

An individual amyloplast is illustrated in the electron micrograph of Plate 6E. Plastid protein is retained with the starch grains when they are released by mechanical shearing from dehydrated pollen (Plate 6B). Germinable pollen grains of rye contain 1500–2000 amyloplasts (Plate 6B, F).

Mitochondria

Bodies presumed to be mitochondria are indicated in Plate 7A, C, electron micrographs from a grain fixed from the dehydrated state with $KMnO_4$ post-staining. The membranes are scarcely discernible.

Lipids

The lipids of the mature grains are mostly dispersed. Irregular aggregates occur at the periphery of the vegetative cell (Plate 7A), and between the polysaccharide particles (Plate 7B), but there are no large globuli in the ungerminated grain comparable with those seen after germination (Plate 13B). The lipid distribution seen in electron micrographs is in accord with that observed following scarlet R staining, so far as the resolution limit of the optical microscope allows a judgment to be made.

Membranes

The protoplast of the vegetative cell of the mature, dehydrated pollen grain is singularly free from distinguishable membranes, even with staining procedures likely to emphasize any that may be present (Plate 7A, B). A continuous plasmalemma is absent, and the boundary of the protoplast is marked by electron opaque masses with the staining properties of lipid (Plate 7A). The envelopes of neither amyloplast nor mitochondria are resolved with clarity, nor can profiles of the endoplasmic reticulum be seen in the crowded cytoplasm. The same processing and staining procedures reveal membranes in the vegetative cell shortly after hydration, showing that the deficiency is not due to the preparation procedure. Evidently the membranes do not have a clear structural identity in the desiccated protoplast.

Ribosomes

Ribosomes cannot be identified in the dehydrated vegetative cells. Azure B and other RNA stains stain the protoplast generally but weakly.

Polysaccharide particles

After the amyloplasts, the most conspicuous component in the cytoplasm of the vegetative cell of ungerminated pollen is a dense population of more or less uniform spherical particles with diameters ranging up to 0·6 μm, and so visible readily enough with the optical microscope. These particles are the counterparts of the 'vesicles' described from the pollen grains and growing pollen tubes of species of various other families (e.g. *Lilium*, Dashek and Rosen, 1966, Rosen, 1968, van der Woude, Morré and Bracker, 1971; *Lycopersicum*, de Nettancourt *et al.*, 1973; *Oenothera*, Dickinson and Lawson, 1975), and known to be implicated in the growth of the pollen tube wall.

In the pollen of rye fixed from the dehydrated state, the particles, while clearly discrete entities, do not have a discernible bounding membrane, although the surface is occasionally demarcated by lipid aggregates (Plate 7 B, C). The description 'vesicle' seems inappropriate for this condition, and the noncommittal term 'particle' is preferable. Since the content is largely polysaccharide, the abbreviation *P-particle* is used in this paper.

The P-particles show a characteristic internal structure consisting of coarse granulations (the 'flecking' of van der Woude, Morré and Bracker, 1971). Occasionally the granulations show a vague fibrillar structure suggesting that each represents a tightly packed nodule of fibrils. The P-particles are concentrated mostly in the hemisphere of the vegetative cell opposite to the aperture – segregated, that is, from the starch. In macerates prepared from desiccated pollen, the P-particle-containing cytoplasm stains for protein, and is PAS-reactive (Plate 6C). Particles from disrupted grains processed for electron microscopy are strikingly heterogeneous in appearance (Plate 17A). A proportion shows the coarsely granulate content of those of the vegetative cell when intact. Others reveal a range of internal structure from loose granulation to a more or less uniform fibrillar aspect. Such a range might be interpreted as meaning that the particles enlarge by hydration on release, the tightly packed fibrils progressively dispersing from the nodes until they are uniformly distributed throughout.

Hydration and germination

Dimensional changes

The dimensional changes undergone by rye pollen during hydration on Medium A are summarized in Table 4. The grain as shed is not fully dehydrated, and under natural conditions in the field further desiccation will normally follow, the volume change being accommodated by pleating of the exine (Plate 8 A, B).

TABLE 4. *Volume changes and water flux during hydration and germination of the pollen of* Secale cereale *cv. Rheidol on Medium A*

	Pollen dimensions		
	Maximum diameter (μm)	Minimum diameter (μm)	Volume[1] (cm^3)
As shed	55.73 ± 0.8	45.44 ± 0.6	6.77×10^{-8}
Fully hydrated	70.59 ± 1.3	55.58 ± 0.7	13.16×10^{-8}
Tube emerged	63.53 ± 1.2	49.70 ± 0.9	9.51×10^{-8}

Water flux during hydration[2]		
Total (cm^3)	Rate ($cm^3\ s^{-1}$)	Per unit area ($cm^3\ cm^{-2}\ s^{-1}$)
6.38×10^{-8}	5.31×10^{-10}	2.71×10^{-5}

[1] Calculated as a sphere assuming a diameter equal to the mean of the maximum and minimum measured dimensions.

[2] The calculations of rate are based upon a hydration time of 2 min and movement through a meniscus of 50 μm diameter.

Time sequence

The time-lapse photomicrograph sequence of Plate 8c shows the detailed behaviour of a grain upon the stigma, illustrating the exudation phase noted by Watanabe (1955), and the later resorption of the exudate before the emergence of the tube. The times to exudation and germination in self- and cross-pollinations of rye are given in Table 5, and some comparisons with other conditions appear in Fig. 6. The results for rye show the importance of the substratum in controlling the rate of hydration and the time to germination. Rye pollen is evidently incapable of abstracting water as rapidly from the stigma of *Dactylis glomerata* as it is from the rye stigma, and germination is delayed accordingly. After the slower uptake of water from a medium containing 0·6 M sucrose through a single Visking dialysis film, the stage of germination was not reached at all. On the other hand, the optimum Medium A permitted a slightly more rapid germination than did the stigma, and germination was advanced accordingly.

Estimates of water flux are given in Table 4 for rye pollen hydrating on Medium A. These are based on the average attained over the whole period of hydration, taking no account of the interpolated exudation phase. The period assumed is 2 min, likely to be an overestimate; accordingly the values for the rate of movement are on the conservative side. The orientation of the pollen grain on the stigma makes little difference to the rate of hydration, indicating that water uptake takes place freely through any part of the surface. The inference from this is that the passage is through the micropores of the tectum in a grain oriented like that in Plate 8c.

TABLE 5. *Times from pollen contact to the formation of the meniscus, exudation and germination of pollen of* Secale cereale. *Single-grain, manual pollinations. Including data from Shivanna, Heslop-Harrison and Heslop-Harrison (1978)*

Pollination	Meniscus appears (s)	Exudate appears (s)	Germination (s)
Compatible	21·7 ± 1·6	36·2 ± 2·6	75·7 ± 5·9
Incompatible	23·5 ± 1·6	43·5 ± 2·9	76·4 ± 9·1

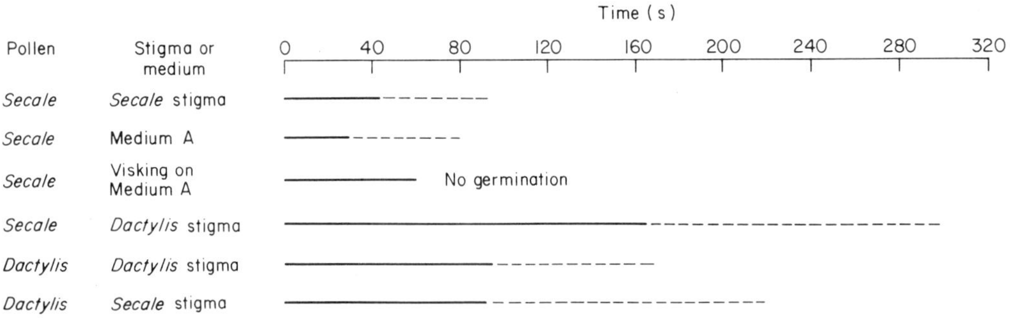

FIG. 6. Times to exudation (solid line) and to germination (interrupted line) in pollen of *Secale cereale* and *Dactylis glomerata* on various substrata.

Loss of germinability during drying

The germinability of rye pollen held in laboratory atmosphere (23 °C, 50 per cent relative humidity) for various periods after shedding is shown in Fig. 7. The proportion of grains germinating fell after 2 h, and pollen kept in open dishes for 9 h was incapable of germinating. The grains then have the appearance seen in Plate 8A, with an estimated water content of 10–15 per cent.

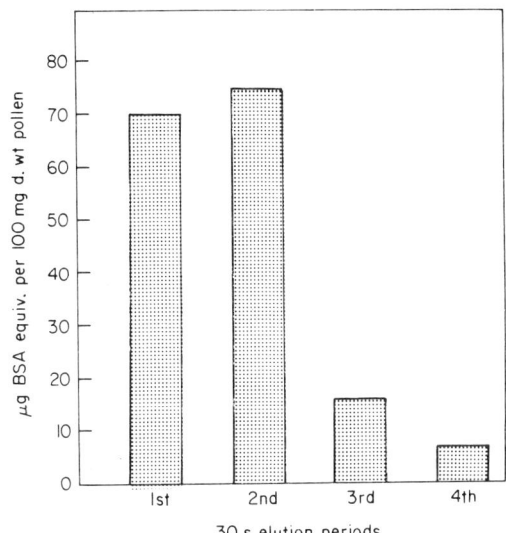

FIG. 7. Protein release from the pollen of *Secale cereale* into 0·4 M sucrose during four successive 30 s elution periods.

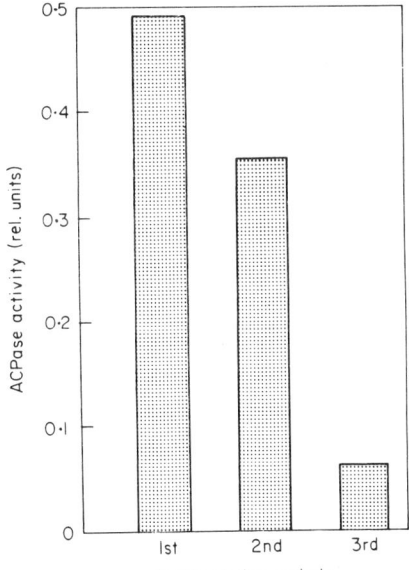

FIG. 8. Acid phosphatase release from the pollen of *Secale cereale* during three successive 2 min elution periods.

Carbohydrate emissions

The release of soluble carbohydrate during three successive 1 min elution periods is shown in Fig. 8. Evidently the bulk is lost very rapidly in the conditions of this experiment. On the basis of R_F and colour reactions following thin-layer chromatography, sucrose was identified as the principal sugar, with appreciable quantities of glucose and arabinose. Galactose was tentatively identified in trace amounts, but not xylose. Two other components, possibly oligosaccharides but as yet unidentified, were present on the chromatograms.

Protein emissions

During the exudation period, fluid is lost from the apertural site, and also over the whole surface of the exine (Plate 8 C). The presence of proteins in the exudation is readily demonstrated by simple staining procedures (Heslop-Harrison *et al.*, 1975), which can, nevertheless, give resolution adequate to distinguish apertural and intine emissions from grass pollens (Plate 8 D, E, F).

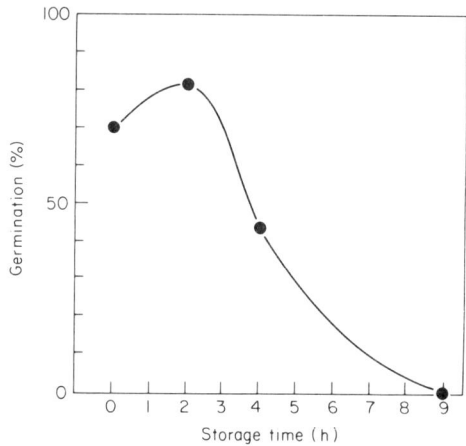

FIG. 9. Germinability of pollen of *Secale cereale* on Medium A after being held for various periods of time at 23 °C and 50 per cent relative humidity.

The time course of protein emission into 0·4 M sucrose is shown in Fig. 9, which covers four successive 30 s elution periods. Almost 70 per cent of the protein is lost within the first minute. This would, of course, cover the period of exudation from rye pollen hydrating on the stigma (Plate 8 C), when, presumably, the greater part of the protein release onto the stigma surface takes place.

The elution of the pollen into 0·4 M sucrose ensures that osmotic bursting of the grains is kept to a minimum. Nevertheless, the percentage of grains giving the fluorochromatic reaction fell somewhat from the first to the last elution, indicating that cytoplasmic proteins from disrupted grains may have contributed to the measured protein yield. Such a contribution must be quite small, however, and the pattern of protein release observed probably reflects quite closely that occurring naturally in the course of hydration following pollination.

Separation of the proteins of short-term eluates of rye pollen by iso-electric focusing on polyacrylamide gels reveals some 30 components (Heslop-Harrison, 1978), including three glycoproteins. Isoenzymes of acid phosphatase and non-specific esterase are present in the eluates, presumably derived from the intine. The release of acid phosphatase in

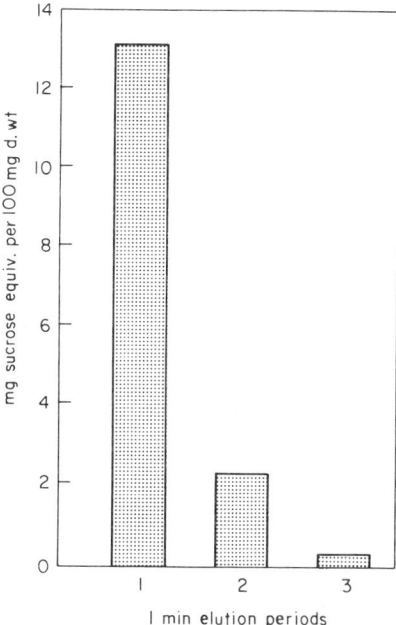

FIG. 10. Release of soluble carbohydrate from the pollen of *Secale cereale* during three successive 1 min elution periods.

three successive 2 min elution periods is shown in Fig. 10. Again it may be seen that the loss is principally during the first period, spanning the interval when, on the stigma, the grains would be exuding.

The exudates passing onto the pollen surface and into the meniscus between the grain and stigma papilla are readily leached away by water, and the protein fractions cannot be fixed readily for electron microscopy with the usual fluid fixatives. Fixation with osmium tetroxide vapour stabilizes the exudate to some extent, and preparations of pollen grains of rye made in this way are illustrated in Plate 2. The exudate is seen as a halo enveloping the exine. Material of similar staining properties is sometimes seen in the exine cavities, and continuities through the micropores of the nexine suggest that the intine is the principal source.

Protein uptake during hydration

The capacity of the unhydrated but viable grains to take up FITC-labelled proteins was tested with the pollen of *Hordeum bulbosum*. Labelled pollen- and stigma-proteins were found to enter some grains during the first minute after irrigation, but the behaviour was irregular. Larger grains in the pollen samples showed only slight fluorescence at the characteristic FITC wavelength, and this was mostly at the periphery of the vegetative cell. The smaller grains showed a more general distribution of fluorescence, suggesting passage throughout the protoplast. Pre-hydrated but ungerminated pollen showed little or no uptake of any labelled protein. The results must be considered inconclusive.

The cytoplasm during germination

During the first 1·5–4 min of hydration the cytoplasm of the vegetative cell undergoes extensive re-organization. By the time the tube tip has emerged a distance of 20 μm, streaming can be observed in the tube, and within 5 min, movements of the protoplasm can be seen elsewhere in the cell. Because of the rapidity of the changes and the variation between grains it is not feasible to make an accurately timed sequence of fixations for electron microscopy. However, various morphological markers are available, including the bulging of the apertural intine and the dispersal of the Zwischenkörper, changes that are followed shortly by vacuolation and the passage of amyloplasts into the emerging tube. By reference to these markers, a sequence can be constructed from pollen populations fixed during hydration and germination. The following account is based upon such a sequence; the changes described occur in 3–4 min from the beginning of hydration.

Lipids. Irregular lipid aggregates at the surface of the protoplast and around the P-particles (Plate 7A, B) progressively disappear, and globuli appear in the cytoplasm (Plate 14A, B, C).

Mitochondria. The appearance of the mitochondria in the ungerminated grain, characterized by the poor definition of the envelope and internal membranes (Plate 7C) is lost, and the mitochondria assume an aspect more characteristic of those of an active somatic cell (Plate 9B).

Amyloplasts. With the lifting of the operculum and the bulging of the apertural intine, the slightly elongated starch grains reorient from a random distribution (Plate 3A) to a polarized one (Plate 3B), suggesting a flow pattern focused on the aperture. The first starch enters the tube in 4–6 min from the onset of hydration. There is no indication of the erosion of the starch at this time.

Membranes. A continuous plasmalemma cannot be resolved in the ungerminated grain, and the cytoplasmic membranes are ill-defined or absent (Plate 7A). By the time the first vacuoles appear, the plasmalemma, tonoplast and membranes of the endoplasmic reticulum can be resolved once more (Plate 9B). The membranes of the endoplasmic reticulum lack ribosomes.

P-particles. The coarsely granulate or flecked appearance of the contents is retained during early hydration and vacuolation (Plate 9A, B). Thereafter the packing of the contents of the particles loosens, and many show a fibrillar component with a distribution suggesting a derivation of the fibrils from the denser foci of the dehydrated particle (Plate 14A, B, C). By this stage bounding membranes can be resolved. These are, however, usually discontinuous; few particles appear to be wholly invested, and frequently they appear to be coalescing.

The fate of the Zwischenkörper

Early in the course of hydration of a viable grain, the thin sporopollenin membrane subtending the operculum at the germination aperture is ruptured, and the underlying Zwischenkörper is gelatinized. Alcian blue stainability is substantially diluted at the apertural site, and the gel produces a prominence that displaces the operculum, either pushing it aside, or carrying it out away from the exine (Plate 12C). The stainability of the gel is low, but its presence is readily demonstrated by the exclusion of indian-ink particles from the vicinity of the aperture in grains hydrating *in vitro* (Plate 10A). It is during this interval that the emission of proteins from the apertural intine begins (Plate 8E).

In a liquid medium that does not support complete germination, the emerging tip tends to stabilize at this point, and in dilute indian ink the gelatinized cap then accumulates carbon particles (Plate 10B). On a semi-solid germination medium, the gel is displaced by the tip of the tube and is ultimately lost, presumably by ablation into the medium.

Organization of the emerging tube tip

The tube tip develops first as a papilla on the intine. This arises as a zone of weakness immediately below the shaft of the aperture. Ghost intines from grains at this early stage of germination often show a perforation at this site (Plate 11 c), presumably because the vigorous chemical extractions involved in preparing the ghosts disperse the softened wall. The weakened area is always in close register with the site formerly occupied by the overlying Zwischenkörper; the thickened intine below the exine annulus of the aperture margin is not affected. After this first stage the emerging papilla acquires greater mechanical strength (Plate 12 A, B).

The electron micrograph of Plate 13 A is of an oblique section of the tip region during the emergence of the tube. It shows that the tip has a microfibrillar component; the microfibrils give the impression of being short, and have no preferred orientation. No well-defined plasmalemma is present, and the interface between wall and cytoplasm is occupied by numerous bodies with microfibrillar content, the derivatives of the P-particles. The wall in the tip region is not noticeably stratified. The whole thickness is PAS-positive, and in grains germinating under normal conditions no callose is evident at this stage.

Pseudo-germination of dead pollen

Inviable stored pollen hydrates rapidly on Medium A, reaching a volume comparable with that of viable pollen at the time of tube emergence. In many grains, the operculum then lifts, and the intine emerges, much as in a normal germination. Thereafter the intine balloons outwards, ultimately bursting (Plate 12 D). Evidently matric imbibition alone is sufficient to dilate the grain and create the pressure initiating the emergence of the tube tip. The subsequent isodiametric swelling of the apertural intine shows that the enlargement is by plastic expansion and does not involve growth of the wall.

Cytoplasmic zonation during tube emergence and early growth

In rye, the tubes grow towards the stigma and over its surface at an intial rate of $1·3–1·6\ \mu\text{m s}^{-1}$ (Shivanna, Heslop-Harrison and Heslop-Harrison, 1978). This rate is maintained for the first 30–50 μm of growth, and while it is progressing, a zonation of the cytoplasm is established. Immediately behind the extending tip, the main cytoplasmic components are the P-particles, usually rather tightly packed. Few membrane profiles are evident, and no organelles. A sparse population of smaller, membrane-bounded vesicles is present, distinguishable from the P-particles by their homogeneous contents. This distal zone extends for 25–30 μm. Mitochondria and membranes appear in the succeeding zone, and proximal to this again, amyloplasts are also present. No dictyosomes have been observed in the tube during the early period of extension.

The growth is accompanied by vigorous cyclosis. In the zone immediately behind the extending tube tip, particles, presumed to be the P-particles visible in electron micrographs, undergo random movements of considerable amplitude, but show no directional flow pattern. Behind this zone, there is an apparent net flow of particles towards the tip, and behind this again, a form of irregular 'fountain' streaming can be observed. Amyloplasts and inclusions of the dimensions of mitochondria move towards the apex of the tube in a central core and are then arrested, or are caught in a reverse current and carried away from the apex again. The micrograph of Plate 10 E, from a time-lapse sequence made with electronic flash illumination, provides a 'frozen' image of the streaming protoplast; it suggests that the zonation of tube contents seen in fixed preparations reflects a sorting out of the cell constituents determined by the pattern of cyclosis.

Stratification of the older wall

An inner lining of callose is laid down in the older wall, the callosic zone extending in pace with tube growth and terminating 10–15 μm behind the tip. The resultant stratification is seen in the electron micrographs of Plates 12E and 13.

The *outer sheath* of the older wall is *c.* 3 μm in thickness, has a microfibrillar component, and is PAS-reactive. Electron micrographs suggest that the microfibrils are oriented circumferentially.

The *inner lining*, *c.* 2·5 μm in thickness, is homogeneous in electron micrographs. It is not PAS-reactive, but gives strong fluorescence with decolorized aniline blue.

The two layers are distinct from each other, and indeed separate one from the other in the older tube (Plates 13B and 17C).

P-particles and wall growth

Different aspects of P-particles released from disrupted ungerminated pollen of rye are illustrated in Plate 17A, and according to the interpretation given earlier these may reflect different degrees of relaxation of the polysaccharide content on hydration, the most fully relaxed presenting a more or less uniform microfibrillar aspect. A similar range of states can be seen in the P-particles of pollen following hydration and the beginning of tube growth (Plates 13B, C and 14A–C). The larger aggregates (Plate 13B) apparently arise from the coalescence of several individual particles, as judged from the inclusion of membrane fragments.

In the growing tube, microfibrillar masses with the appearance of the P-particle content are seen adjacent to and in continuity with the wall at the tip region. In the callosic zone similar profiles appear, with the P-particles apparently fusing with, and being taken into, the callose inner lining (Plate 13C). So far as can be established from the static evidence of electron microscopy, it seems that the P-particles are the precursors of the wall, constituting the principal means whereby the stored polysaccharide of the grain is moved to the extending regions of the tube and incorporated into the thickening cellulosic and callosic layers.

Sources of the P-particles: starch and lipid utilization

Earlier authors have for the most part assumed that the P-particles of pollen tubes are derived from the activity of the Golgi system, the vesicles arising in the characteristic fashion from dictyosomes, and accumulating polysaccharide content as they mature. None of the published ultrastructural accounts of pollen development in the grasses carries the sequence through to the mature, dehydrated grain, so direct evidence of the origins of the large population of P-particles in the vegetative cell is not available. However, several published electron micrographs do show substantial populations of dictyosomes in the maturing grains after the completion of intine thickening (e.g. de Vries and Ie, 1970, figs 6–8), and it is entirely probable that these are concerned with the production of the P-particles.

Certainly the circumstantial evidence points strongly to the conclusion that the stored P-particles of the grain are the main immediate source of wall precursor materials during germination and early tube growth. Ultimately, however, the other reserves of the vegetative cell are used, including lipids and starch, and the likelihood is that they also contribute to the tube wall. A possible route for starch would be through degradation into glucose and reassembly into the appropriate wall precursors through the intermediacy of the Golgi system, the dictyosome vesicles then differentiating into further populations of P-particles, for transport into the wall. However, dictyosomes are extremely rare in the cytoplasm of the vegetative cell of the rye pollen grain, and in no instance has a secure indication of dictyosome activity in the formation of P-particles been encountered at this stage of development.

On the other hand, direct associations between starch grains in the course of degradation and bodies with a microfibrillar content comparable with that of P-particles are invariably to be seen in the active period of tube growth. Degrading starch in a germinating grain is illustrated in the optical micrograph of Plate 6D, and associations with P-particles in the electron micrographs of Plate 14. At this time the amyloplast envelope is discontinuous or absent altogether, and the associated P-particles are similarly without continuous bounding membranes, although short profiles are usually present. Mitochondria are frequent in zones of the cytoplasm where the starch is undergoing degradation.

The lipid reserves of the grain are also mobilized during the growth of the tube, and it may be significant that lipid globuli are also seen in direct association with simple or compound P-particles (Plate 14c).

Dissolution of the intine

The protein inclusions of the intine, inconspicuous in the ungerminated grain, become more prominent as hydration proceeds, appearing in electron micrographs in irregular tubular or vesicular conformations (Plate 3c). As this change takes place, the microfibrillar component of the intine also becomes more conspicuous, and the layer then undergoes progressive dissolution. A characteristic of this stage is the presence of microfibrillar bodies, variable in size but strikingly similar in other respects to the P-particles, associated with the inner face of the intine (Plate 16A). Detached bodies with the same content are present in the protoplast as it withdraws from the grain (Plate 15A, B). The final withdrawal of the cytoplasm into the tube leaves the former site of the intine occupied by a loose reticulum of fibrils (Plate 16B), or occasionally, a sparse population of residual vesicles (Plate 16c). The relationship of the intine residues with the proximal end of the pollen tube is shown in Plate 15A, c. Plate 15A shows that the exine annulus surrounding the aperture is the focus of a funnel shaped collar of wall material, which grades into the residual intine within the grain and is continuous with the tube at the exit. At higher magnification, it can be seen that the microfibrils of the wall, randomly disposed within, are oriented in the direction of tube extension in the shaft of the aperture (Plate 15c).

Protein and enzyme emission by the emerging tube

The proteins released from the aperture and through the exine are probably derived from intine sites during the early period of hydration. This emission usually declines before the emergence of the tube tip from a germinating grain. After the dispersal of the Zwischenkörper, however, and with the establishment of zonation in the tube tip, a second period of emission begins, from the tip zone itself (Plate 10c). This emission at first forms a cap over the apex, and then later appears mainly to be from the flanks of the tube immediately behind the tip (Plate 10D), extending back to the region where the deposition of the callosic inner lining of the tube begins. The release of proteins from the growing apex continues thereafter, so that the tube leaves a trail over the medium (Plate 10F). The standard localization methods reveal the presence of acid phosphatase and non-specific esterase in the early exudate, and the substrate film technique using apple pectin as substrate shows that pectinase activity is present in both apertural and exine exudates (Plate 18A). Presumably these are members of a family of enzymes secreted by the extending tube.

While the source of the secreted proteins cannot be specified with any certainty, it is probably significant that they are released from that part of the growing tube where the heterogeneous population of single-membrane bounded vesicles seen in Plate 13c, D occurs dispersed among the P-particles. These vesicles are often seen apposed to the wall, suggesting a secretory function.

After the establishment of the characteristic zonation in the growing tube, the organelle-free tip region shows moderate affinity for protein stains, and also a general – although light – affinity for RNA stains. Ribosomes cannot be distinguished with certainty in the tip region in electron micrographs of preparations post-stained by standard methods, nor can membranes of the endoplasmic reticulum. This may be due to the extraordinarily dense packing of the P-particles.

Light micrographs of living tubes show marked longitudinal striation in the tube (Plate 10E). This is lost on fixation, and the cause is not apparent in electron micrographs.

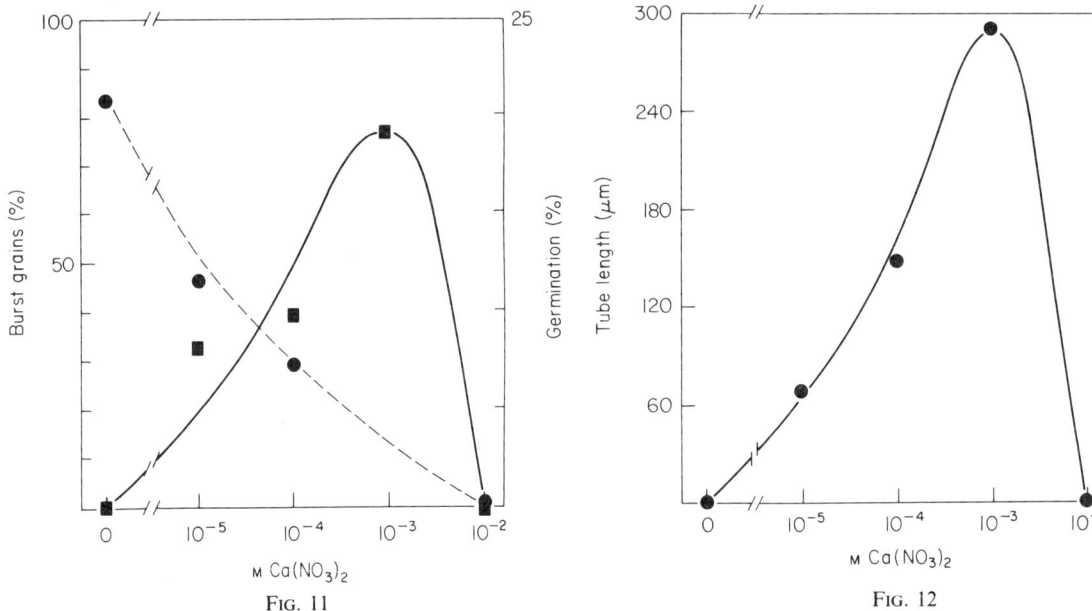

FIG. 11. Effect of calcium concentration on germination of the pollen of *Hordeum bulbosum*. The medium contained 1 per cent agarose, 0·6 M sucrose, 10^{-3} M boric acid and the concentrations of calcium nitrate indicated. The scoring was done 1 h after the sowing of the pollen. ●---●, Proportion of grains bursting; ■——■, proportion germinating.

FIG. 12. Tube lengths attained at the time of scoring in the group of grains that did germinate in each sample.

Calcium and pollen germinability

The effects of calcium concentration in the medium on germination and tube growth in pollen of *Hordeum bulbosum* are shown in Figs 11 and 12. The proportion of grains bursting immediately after transfer to the medium falls with increasing calcium concentration. The proportion germinating reaches a maximum at 10^{-3} M $Ca(NO_3)_2$ and falls steeply thereafter. The maximum tube length attained follows the same trend.

At the highest concentration, 10^{-2} Ca^{2+}, viable pollen hydrates readily, and passes through the exudation and resorption phases. However, the Zwischenkörper does not gelatinize. The grains continue to dilate (Fig. 14), reaching volumes considerably

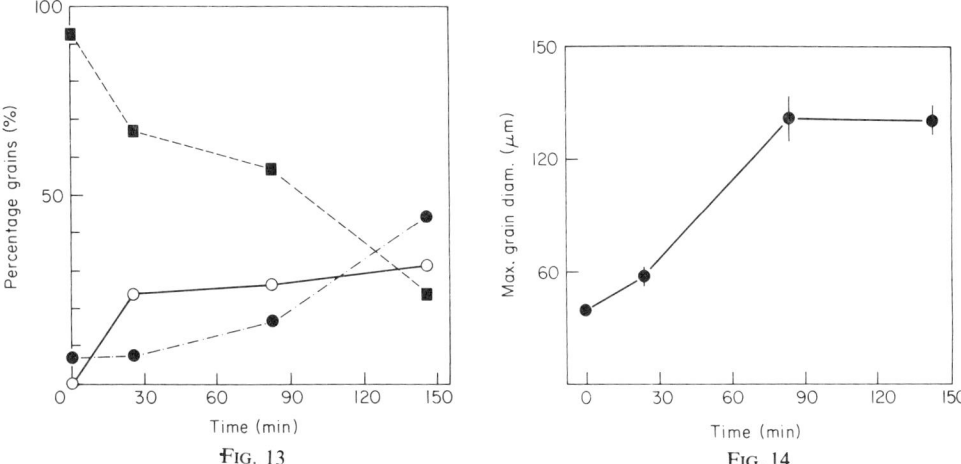

FIG. 13. Changes in starch content in pollen of *Hordeum bulbosum* during incubation at 24 °C on Medium B (10^{-2} M calcium nitrate). The grains were scored in three categories: ■---■, with intact starch; ○----○, with eroding starch (cf. Plate 6B showing a late stage in starch erosion in rye); ●—·—●, without starch. Starch stained with I-KI.

FIG. 14. Volume changes in the pollen of *Hordeum bulbosum* during incubation at 24 °C on Medium B.

greater than the maximum before germination would normally occur. Dilated grains of *H. bulbosum* on Medium B are illustrated in Plate 18B.

Vacuoles appear in the vegetative cells of calcium-blocked grains in 4–6 min, and protoplasmic streaming then begins. As the time lapse sequence of blocked pollen of rye in Plate 18C shows, the movements are vigorous, and they continue for 3–4 h. During this time the storage starch of the grain is progressively eroded (Fig. 13, for *H. bulbosum*) and the lipid globuli disappear from the vegetative cell. The principal constituent of the cytoplasm at the end of the period is a population of P-particles.

A proportion of grains of *H. bulbosum*, while expanding during the early period of hydration, fails to continue dilation (Plate 18B). These grains are permeable to dyes, suggesting that the plasmalemma is leaky. They show no cyclosis, and the starch content is not eroded during further incubation.

DISCUSSION
Hydration and emission

The sequence recorded by Watanabe (1955) as typifying a successful pollination in the grasses (Fig. 5) includes hydration, exudation, resorption and germination phases. These events may be interpreted on the assumption that the behaviour of the grain as an osmotic system changes during hydration in consequence of alterations in the properties of the membranes of the vegetative cell (Heslop-Harrison, 1979a). The argument is based upon the proposition that the membranes of the dehydrated grain are ineffective as osmotic barriers, the condition surmised for the membranes of desiccated seeds (Simon, 1974) and suggested by the electron microscopic evidence, but that they undergo a transition during hydration into a normal configuration, and in doing so regain the capacity for controlling the movement of solutes. The sequence envisaged runs as follows.

Following the capture of the grain, water will flow into it from the stigma papilla as long as the water potential of the stigma exceeds (i.e. is less negative than) that of the grain, i.e. while $\psi_{stig} > \psi_{poll}$. Assuming that ψ_{stig} remains constant, the flow will be governed by ψ_{poll}. The components contributing to ψ_{poll} are ψ_s, attributable to the solutes on or in the grain, ψ_m the matric potential related to the colloids of the cytoplasm, and ψ_p, wall potential. If the grain is initially desiccated the wall may be under negative tension, so that ψ_p will have a negative value.

Endosmosis of water will dilate the grain and hydrate the colloids of the cytoplasm until ψ_p reaches zero and then assumes positive values. While the membranes of the vegetative cell remain ineffective as osmotic barriers, the wall will not constrain the movement of solutes, but it will retain the structurally anchored colloids of the cytoplasm. Accordingly, expansion will continue until $-\psi_p = \psi_m$, when wall pressure balances that generated by the cytoplasmic colloids, at which point the component of water flux attributable to ψ_m will fall to zero. Nevertheless, the passage of water from the stigma will not cease at this time because the solutes associated with the grain are not constrained by the wall; the flow will continue while $\psi_{stig} > \psi_s$. However, since the maximum water capacity of the grain has been attained when $-\psi_p = \psi_m$, this further passage must result in exudation.

These considerations indicate how the movement of water from the stigma will at first serve to hydrate and dilate the grain, and how this phase will be followed by a second one, when water and solutes are lost by exudation from the germination aperture and through the micropores of the non-apertural exine.

The second phase will be followed by a third, when the membranes of the cell are reconstituted. The vegetative cell of the pollen grain will then be translated into an osmotic system of the ideal plant cell type, bounded by a semi-permeable membrane contained within an elastic wall. Further water flux will be determined by the solutes contributing to ψ_s and changes in ψ_p. The further depression of the water potential of the grain by the containment of the solutes of the vegetative cell within an effective plasmalemma will lead to the endosmosis of water and still greater dilation until the balance is once again restored. Furthermore, water will move back into the grain from the exudate if the water potential of the exudate were greater than that of the vegetative cell.

With the completion of this sequence, a considerable hydrostatic pressure will be built up in the grain, to be relieved either by bursting, or by the emergence of the apertural intine and the beginning of tube growth.

The general interpretation just outlined is likely to apply to all pollens that are dispersed in a partly dehydrated state. It accounts for the unexpected spectacle of exudation during the course of hydration – exudation which is indeed bound to occur wherever the recovery of the membranes of the grains is delayed until after the first dilation of the grain has brought wall and matric pressure into balance. It accounts also for the massive loss of sugars from hydrating pollen; inevitable if there is no containment by the membranes. It seems, further, that in the case of angiosperm pollens the early sequence of events has been pressed into a positive biological use, for it is during the brief interpolated period of efflux from the grain that materials from the pollen are transferred to the stigma surface. For grass pollens, these include the intine-held proteins which pass out not only from the aperture site but through the micropores over the whole surface of the exine (Plates 2A and 8D). This transfer has obvious significance in the interactions between the pollen and the stigma, including those associated with the incompatibility system (Heslop-Harrison, Heslop-Harrison and Barber, 1975).

The proposition that the membranes of the dehydrated pollen are 'leaky' carries another significant implication, namely that at the time of first contact with the stigma there may be no effective barrier to the penetration of stigma surface materials. The initial rapid inflow may well convey stigma materials not only up to, but through, the plasmalemma, another possibility of great import for the pollen stigma interaction. The attempt to demonstrate entry of FITC-labelled proteins during the early hydration of rye pollen had no satisfactory outcome, but further experiments along the same lines are clearly needed to discover precisely how much of a barrier the membranes of partly dehydrated pollens actually are.

Germination

In the normal course of events, germination follows immediately upon the resorption phase (Fig. 5; Plate 8c), when the foregoing theoretical treatment would suggest a rising hydrostatic pressure in the grain. It is now clear that in grasses the Zwischenkörper plays an essential part at this time. The principal constituents of the Zwischenkörper are pectins, forming a consolidated crust over the apertural intine in the ungerminated grain, a crust protected above by the sporopollenin lamella supporting the operculum. During the earliest stages of imbibition this lamella is disrupted and the Zwischenkörper forms a gel, pushing out the operculum and so opening the aperture. With the hydration of the intine, the enzymes borne in the intine inclusions are released, and the cytochemical evidence shows that they diffuse through the residual material of the Zwischenkörper. The presence of pectinase suggests that these residues will be further degraded by the intine emissions, while the fabric of the intine itself will be loosened at the apertural site, to the extent that its strength is determined by matrix polysaccharides. All of these events occur in both viable pollen and pollen that has lost its germination capacity through storage. They must therefore depend on the operation of in-built systems that do not require metabolic activity in the vegetative cell.

The divergence follows during the next stage of tube emergence. In inviable grains, hydration in a suitable medium leads to the extrusion of the intine at the aperture; development either ceases at this point, or the extruded wall bursts with the release of cell contents. In the viable grain, a zonation is quickly developed, generally within 3 min from the beginning of hydration. This is associated with a transition from simple inflation of the emerging intine to a tubular growth pattern with the wall microfibrils oriented in an annular fashion. As Plates 10c and 12a show, this is achieved in under 10 μm of growth and in a time interval of less than half a minute. Whether the new growth pattern is first established by the redistribution of existing intine material or by the oriented deposition of new microfibrils, it evidently requires the participation of a viable protoplast.

The successful passage from the early expansion of the grain, through the stage of opening of the aperture and intine emergence to the establishment of the tip-growing tube must depend upon the maintenance of a critical balance between several factors, and most notably between the mechanical strength of the emerging intine and the osmotic relationships of grain content and environment.

The mechanical properties of the intine at the aperture are determined by a number of structural components, the main ones being the Zwischenkörper, the microfibrillar framework of the intine itself, and the hemicellulosic and pectic materials of the matrix encasing this framework. The importance of water in determining the properties of the Zwischenkörper has been emphasized. Hydration will also alter the state of the matrix materials, increasing the plasticity, and by weakening the hydrogen bonding between microfibrils and matrix materials will enhance the possibility for redistribution and reorientation. The effect of supra-optimal calcium concentrations in blocking germination of grass pollens presumably arises from the bonding and rigidification of the pectins of the Zwischenkörper and the matrix materials of the intine. But the strength of the matrix materials will also be modified by the activity of the wall-held carbohydrases, certainly with an effect on the plasticity of the apertural intine. In the case of the grasses, pectinase would seem to be the most important factor, and it is therefore significant that it is released from the grain during early hydration. In other species, a somewhat similar function to that attributed here to the Zwischenkörper could perhaps be discharged by callose. Roggen and Stanley (1969) showed that callose is formed at the apertural sites of pear pollen, and that the addition of β-1,3-glucanase to the medium accelerates germination. Significantly, both β-1,3- and β-1,4-glucanases are released immediately after hydration of pear pollen (Stanley and Linskens, 1974). However, the sites of callose in the apertural regions of pear pollen is not precisely known, and the account given by Roggen and Stanley (1969) suggests that the main deposition may occur *after* germination. If so, the effect of β-1,3-glucanase supplied exogenously might be secondary, and the earlier control may be akin to that postulated here for the grasses.

The importance of osmotic balance has been appreciated since the earliest attempts to germinate pollen in artificial media (Stanley and Linskens, 1974). In general, each species has an optimal range, but it is common knowledge that within this range germinability may vary considerably according to the state of the pollen sample and on other factors in the medium. The observations of Watanabe (1961) point to the possible importance of the starch/sugar balance within individual grains of grass pollen; and this is no more than to be expected since sugars are likely to constitute the principal osmoticum in all pollens (Iwanami, 1959). According to the earlier argument, the restitution of the membranes of the grain during hydration marks a critical point for the transition of the vegetative cell from an open, leaking system to an osmotically contained one; thereafter the solute potential must drive further expansion and eventually the extrusion of the apertural intine and the enlargement of the tube tip. The important balance is between the concentration of osmoticum in the medium, and that contributed by the free sugars already present in the grain when the membranes are re-established and enhanced thereafter by the metabolic conversion of starch and other reserves.

We see, then, that the factors determining the capacity of the grain to expand, germinate and produce an extending tube in a given medium will include (*a*) the state of the reserves, (*b*) the condition of the membranes, and (*c*) the subsequent rate of reserve conversion. Factor (*a*) will be related in the main to the metabolism of the parent plant during the final stages of pollen maturation. Factors (*b*) and (*c*) may in part be controlled by the parent, but may also be dependent on metabolic machinery set up under the genetic control of the male gametophyte itself during post-meiotic development in the anther. Evidently environmental factors such as temperature and oxygen tension will impinge mainly on the metabolically dependent element under factor (*c*).

Several observations point to the delicacy of the balance between mechanical and osmotic controls in the germination of grass pollens. In a natural pollination, hydration is not in a fluid medium, but from a 'dry' stigma. The water uptake, although rapid, is thus controlled, and probably self-regulating (Heslop-Harrison, 1976, 1979*b*). *Un*regulated hydration leads to the bursting of the grain. This suggests that the difficulty of germinating grass pollens in fluid media probably arises from the simple fact that the high sugar concentrations necessary to retard the initial ingress of water are too high to permit the emergence of the tube thereafter. For this reason, semi-solid media which simulate to some extent the conditions of the stigma are needed to promote normal germination, as convincingly shown by Bar Shalom and Mattsson (1977) for the Gramineae and other dry stigma families.

As we have noted, the role of calcium in the regulation of germination must be through its effects on the mechanical properties of the Zwischenkörper and the underlying intine. Sub-optimal calcium concentrations increase the proportion of bursting grains (Fig. 11), and supra-optimal concentrations block germination. The proof that this is wholly due to the effect on the wall is given by the observation that the metabolic conversion of starch continues in the blocked grains, which then enlarge to several times the size at which they would normally germinate in consequence of continued endosmosis of water from the medium. The containment of the cell in these circumstances depends upon the effectiveness of the plasmalemma as an osmotic barrier and the mechanical strength of the wall which supports it. Agents that increase the permeability of the cell membrane permit the loss of osmoticum and the elastic recovery of the wall. Grains that do not enlarge on the high-calcium medium have leaky membranes and are thus inviable in any event.

The pollen tube tip and the zonation of the tube contents

As we have seen, the tip of the tube emerges first as a prominence of the intine extruded through the shaft of the germination aperture; true growth ensues when the expansion ceases to be isodiametric and the tubular pattern of extension is adopted.

The first expansion of the intine into the aperture is made possible by a weakening of this layer immediately below the shaft, as intine ghosts prepared at this stage clearly show (Plate 11 C). On the basis of the earlier discussion, it might seem sufficient to suppose that this results from changes in the strength and binding capacities of the matrix materials. This must surely be a significant factor; but it is also true that isolated intine ghosts subjected to extraction procedures that might be expected to remove pectins and other matrix materials retain their structural integrity. Moreover, they disperse only tardily in pectinase, and perhaps only because of residual β-1,4-glucanase activity in the enzyme. The implication is that the early weakening of the apertural intine does involve enzymic attack on the microfibrillar component and not only a slackening of the meshwork by hydration and the partial dissolution of matrix materials.

The development of the tubular growth pattern requires that the weakening should be localized, and that thereafter new wall material should be inserted principally at the tip, maintaining thenceforth the delicate relationship between wall weakening, osmotically driven expansion and supply of wall precursors common to all tip growing cells. The transition from disordered expansion to ordered tip growth has not been followed in detail in the present work, but there can be no doubt that it is connected with the setting up of a particular pattern of protoplasmic streaming in the tube and the zonation of the cytoplasmic components. By the time active tip growth has begun, the zonation of Plate 10C, D is established, and this is maintained throughout the whole further course of growth. In general aspect, this zonation is similar to that described for other pollen tubes, but the grasses show important differences.

Excellent accounts are available of the tip zonation in pollen tubes of *Lilium* species, dating from the early work of Rosen and collaborators (reviews: Rosen, 1968, 1971) and including the detailed studies of van der Woude and Morré (1967) and van der Woude, Morré and Bracker (1971). The pioneering paper of Rosen, Gawlik, Dashek and Siegesmund (1964) showed that the tip region is rich in two classes of vesicles, a larger type with a content resembling the wall in staining reactions, corresponding to the P-particles of the present paper, and a smaller type, 10–50 nm in diameter, which may correspond to the smaller vesicles of the grass pollen tube tip (Plate 13C, D). Van der Woude *et al.* (1971) showed that the principal concentration of both classes of vesicles was in the apical 5 μm of the lily tube tip, where smooth-surfaced membranes of the endoplasmic reticulum are also present. The immediate sub-apical zone showed a concentration of mitochondria and active dictyosomes. Although free ribosomes were present throughout the tube, ribosomal endoplasmic reticulum was present mainly in the older regions, where amyloplasts and lipid globuli also occurred. The vesicles corresponding to the P-particles were found to be much less frequent behind the tip zone proper. A generally similar zonation has been described for two dicotyledons, *Petunia hybrida* (Sassen, 1964) and *Lycopersicum peruvianum* (Cresti *et al.*, 1977). The latter authors describe a tip zone 2–4 μm long with two classes of vesicles, the larger again corresponding to the P-particles, in which no other organelles are present. Behind this 'growth' zone, in the sub-apical region of the tube, the cytoplasm was found to be rich in organelles, with mitochondria, amyloplasts, lipid globuli and membranes of the endoplasmic reticulum. This zone contained numerous dictyosomes with the 'active' configuration, associated with – and probably the producers of – a vesicle population.

Comparing the apical zonation seen in *Lilium* and *Lycopersicum* with that of rye as established during the early growth of the tube, three principal differences are seen: (a) the P-particle population in rye tubes is not concentrated in the tip, but occurs throughout the tube and in the grain itself; (b) there is no sub-apical zone of active dictyosomes; indeed dictyosomes, although present in the grain and tube, are obviously very much less frequent, and (c) the endoplasmic reticulum is less conspicuous in the tip region, although present elsewhere, usually as smooth-surfaced plates or tubules.

One can account for these structural differences on the assumption that the economy of the grass pollen grain and tube is geared to a much higher rate of development. Whereas tube growth begins in 2–4 min in rye and attains a rate of $1.5 \mu m \ s^{-1}$, germination does not begin in *Lilium* and *Lycopersicum* until 30–45 min after the beginning of hydration, and the growth rate achieved is considerably lower – no more than $0.25 \mu m \ s^{-1}$ in *Lilium longiflorum* (van der Woude and Morré, 1967), even although the pathway to be travelled by the tube is 10–15 times longer than in rye. Much of the rapid early growth of the tube in the grasses appears to be accomplished through the immediate utilization of the large stock of P-particles already held in the mature pollen grain. These move into the tube and thereafter are distributed throughout, without zonation (Plate 17 B, C). Wall formation is later at the expense of the stored starch and lipid, and ultimately the intine itself is excavated. The fine-structural evidence offered in this paper suggests that P-particles are formed in immediate association with the resource throughout the tube, and in the grain itself until the intine has been finally scavenged. On the other hand, in *Lilium* all accounts indicate that the P-particles originate in the region of dictyosome activity behind the tip, and not in direct association with the reserve materials in the older part of the tube. This suggests that soluble intermediates traverse the extending tube to the dictyosome zone, where the processing is completed.

The ability of lily pollen tubes to incorporate into the wall soluble precursors supplied through the medium is now well established (Kroh and Loewus, 1968), and indeed it seems that lily pollen tubes 'feed' as they traverse the style (Labarca and Loewus, 1972, 1973). The pollen tubes of the grasses may be a great deal more reliant on the reserves present in the grain, and in the case of rye these are evidently more than adequate to support growth throughout the passage of the tube, since starch from the grain is conveyed even into the embryo sac (Vithanage, unpublished).

The pollen grains of grasses are trinucleate, and as shown by Brewbaker (1967) this condition is associated with a number of other physiological differences. In general, trinucleate pollens have considerably higher metabolic rates than do binucleate, and produce more rapidly growing tubes; in this respect the Compositae provide a dicotyledonous match for the Gramineae (Hoekstra and Bruinsma, 1978; Vithanage and Knox, 1977). The findings of Hoekstra and Bruinsma (1978) show that trinucleate pollens have a more highly organized metabolism at the time of dispersal, and that the subsequent development does not require transcription or protein synthesis. In keeping with this conclusion, it has been found in work to be reported more fully elsewhere that germination and tube growth in rye are insensitive to actinomycin D and cycloheximide, indicating that there is no gene transcription in the gametophyte after the maturation of the pollen, nor indeed any further protein synthesis. It seems, then, that the grass pollen grain can be looked upon as a kind of automaton, programmed for the single objective of delivering the male gametes, and already equipped before dispersal with the systems required for germination and tube growth. Metabolically, the gametophyte is concerned primarily with the synthesis of the wall, and the reserves for this are transferred to the vegetative cell during the final maturation in the anther.

Characteristically, trinucleate pollens tend to be short-lived (Brewbaker, 1967), and this has been attributed to their higher metabolic rate. Grass pollens, as we have seen, quickly lose viability in the natural atmospheric conditions in which they are dispersed, but have a more protracted life if stored at high relative humidity. This observation is scarcely in agreement with the view that the life span is related to metabolic rate, since it might be expected that desiccation should retard metabolic activity and so protract the life of the pollen. More attractive is the proposition that the viability of grass pollens is related to the state of the membranes of the vegetative cell, and notably their capacity for recovering normal permeability properties on rehydration. If the partial dehydration before dispersal results in a loss of the lamellar configuration of the membranes and the adoption of a micellar disposition of the membrane lipids (the 'hexagonal phase' discussed by Simon, 1978), then the potential for subsequent germination and normal tube growth will depend on the preservation of the ability of the membranes to re-form bilayers on hydration. The progressive loss of this capacity may account for the fall in germinability in laboratory air shown by rye pollen (Fig. 9). The change may be related to an increasing loss of order in the membrane lipid aggregates, either through random movement or a further decrease in stability with continued desiccation. This interpretation is strongly supported by the observation that the fall in germinability is correlated with a decline in the numbers of grains showing fluorochromasia, an effect seen in other pollens during storage (Heslop-Harrison and Heslop-Harrison, 1970). The fluorochromatic reaction depends on two principal properties of the vegetative cell: (a) the esterase activity necessary to cleave the fluorescein ester and (b) the possession of an intact membrane capable of retaining the released, polar, fluorescein within the cell. Rye pollen rendered incapable of germination in dry storage retains the enzymic competence, but does not show fluorochromasia because the fluorescein leaks into the medium, showing that the membranes have lost their power of retention.

Grass pollens, and seemingly other short-lived trinucleate pollens, appear to lack the stabilizing systems needed to preserve indefinitely the competence for reorganizing the membranes after desiccation. Such systems must be present in pollens, like that of the pines, which do maintain viability over long periods in dry storage. The difference in membrane organization between the different types would clearly repay study. In this connexion one might recall the remarkable observations of Iwanami and Nakamura (1972), who reported that pollen of lily species retained its viability for long periods in storage in organic solvents including acetone, benzene and ether. These solvents might be expected to destroy the membranes of the grain altogether by extracting the lipid components, yet far from this being so, the germinability of dry pollens immersed in them was found to be maintained or even enhanced. Perhaps the explanation is to be sought in the capacity of the solvents to stabilize the desiccated membranes spatially, the lipids being protected in the micellar phase by associated proteins.

ACKNOWLEDGEMENTS

I wish to thank Professor J. P. Cooper, Director of the Welsh Plant Breeding Station, for the use of facilities, and the Curator of the Royal Botanic Gardens, Kew, for providing the material used in the earlier part of the work.

I am grateful to Dr Y. Heslop-Harrison for her invaluable co-operation throughout the study, and also to Drs K. R. Shivanna and H. I. M. V. Vithanage, who contributed data.

Drs J. E. Christensen and H. T. Horner kindly supplied the original of Plate 1 B.

LITERATURE CITED

AHLOOWALIA, B. S., 1973. Germination *in vitro* of rye grass pollen grains. *Euphytica* 22, 575–81.
BAILEY, I. W., 1960. Some useful techniques in the study and interpretation of pollen morphology. *J. Arnold Arboretum* 41, 141–8.
BARKA, T. and ANDERSON, P. J., 1962. Histochemical methods for acid phosphatase using hexazonium pararosanilin as coupler. *J. Histochem. Cytochem.* 10, 741–53.
BARNABAS, B. and RAJKI, E., 1976. Storage of maize (*Zea mays* L.) pollen at -196 °C in liquid N. *Euphytica* 25, 747–52.
BAR-SHALOM, D. and MATTSSON, O., 1977. Mode of hydration as an important factor in the germination of trinucleate pollen grains. *Bot. Tiddskr.* 71, 245–51.
BATYGINA, T. B., 1974. Fertilisation process of cereals. In *Fertilisation in Higher Plants*, ed. H. F. Linskens, 373 pp. North Holland, Amsterdam and Oxford.
BLAIR, R. H. and LOOMIS, W. E., 1941. The germination of maize pollen. *Science, N.Y.* 94, 168–9.
BOUVENG, H. O., 1965. Polysaccharides in pollen. II. The xylogalacturonan from mountain pine (*Pinus mugo* Turr.) pollen. *Acta chem. scand.* 19, 953–67.
BROOKS, J. and SHAW, G., 1971. Recent developments in the chemistry, biochemistry, geochemistry and post-tetrad ontogeny of sporopollenin derived from pollen and spore exines. In *Pollen: Development and Physiology*, ed. J. Heslop-Harrison, 338 pp. Butterworths Scientific Publications, London.
BREWBAKER, J. L., 1967. The distribution and significance of binucleate and trinucleate pollen grains in the angiosperms. *Am. J. Bot.* 54, 1069–83.
BROWN, C. M. and SHANDS, H. L., 1957. Pollen tube growth, fertilisation and early embryo development in *Avena sativa*. *Agron. J.* 49, 286–8.
CHANDRA, S. and BHATNAGAR, S. P., 1974. Reproductive biology of Triticum. II. Pollen germination, pollen tube growth, and its entry into the ovule. *Phytomorphology* 24, 211–7.
CHO, J., 1956. Double fertilisation in *Oryza sativa* L. and development of the endosperm with special reference to the aleurone layer. *Bull. Nat. Inst. Agr. Sci. Tokyo* 6, 61–101.
CHRISTENSEN, J. E. and HORNER, H. T., 1974. Pollen pore development and its spatial orientation during microsporogenesis in the grass *Sorghum bicolor*. *Am. J. Bot.* 61, 604–23.
—— and LERSTEN, N. R., 1972. Pollen wall and tapetal orbicular wall development in *Sorghum bicolor* (Gramineae). *Am. J. Bot.* 59, 43–58.
CRESTI, M., PACINI, E., CIAMPOLINI, F. and SARFATTI, G., 1977. Germination and early tube development *in vitro* of *Lycopersicum peruvianum* pollen: ultrastructural features. *Planta* 136, 239–47.
DARLINGTON, C. D. and LA COUR, L. F., 1960. *The Handling of Chromosomes*, 3rd edn., 248 pp. Allen & Unwin, London.
DASHEK, W. V. and ROSEN, W. G., 1966. Electron microscopical localisation of chemical components in the growth zone of lily pollen tubes. *Protoplasma* 61, 192–204.
DEITCH, A. D., 1955. Microspectrophotometric study of the binding of the anionic dye naphthol yellow S by tissue sections and purified proteins. *Lab. Invest.* 4, 324–51.
DE NETTANCOURT, D., DEVREUX, M., BOZZINI, A., CRESTI, M., PACINI, E. and SARFATTI, G., 1973. Ultrastructural aspects of self-incompatibility mechanism in *Lycopersicum peruvianum* Mill. *J. Cell Sci.* 12, 403–19.
DE VRIES, A. P. and IE, T. S., 1970. Electron microscopy of anther tissues and pollen of male sterile and fertile wheat. (*Triticum aestivum* L.) *Euphytica* 19, 103–20.
DICKINSON, D. B., 1967. Permeability and respiratory properties of germinating pollen. *Physiologia Pl.* 20, 118–27.
DICKINSON, H. G. and LAWSON, J., 1975. Pollen tube growth in the stigma of *Oenothera organensis* following compatible and incompatible intraspecific pollinations. *Proc. R. Soc. Lond.* B 188, 327–44.
ENWRIGHT, J. B., FRYE, F. L. and ATWAL, O. S., 1965. Ribonuclease activity of peripheral leucocytes and serum in rabies-susceptible and rabies-refractory mice. *J. Histochem. Cytochem.* 13, 515–7.
ERDTMAN, G. 1943. *An Introduction to Pollen Analysis*. Chronica Botanica, Waltham, Mass.
—— 1952. *Pollen Morphology and Plant Taxonomy*. Vol. I. Angiosperms. Chronica Botanica, Waltham, Mass.
—— 1960. The acetolysis method. A revised description. *Svensk. Bot. Tidskr.* 54, 561–64.
ESCHRICH, W., 1961. Untersuchungen über den Ab- und Aufbau der Callose. *Z. Bot.* 49, 153–218.
—— and CURRIER, H. B., 1964. Identification of callose by its diachrome and fluorochrome reactions. *Stain Technol.* 39, 303–7.
FAEGRI, K., 1956. Recent trends in palynology. *Bot. Rev.* 22, 639–64.
FERRARI, T. E. and WALLACE, D. H., 1975. Germination of *Brassica* pollen and expression of incompatibility *in vitro*. *Euphytica* 24, 757–65.
GOSS, J. A., 1968. Development, physiology and biochemistry of corn and wheat pollen. *Bot. Rev.* 34, 333–58.
GURR, E. 1965. *The Rational Use of Dyes in Biology*. 422 pp. Leonard Hill, London.
HESLOP-HARRISON, J., 1968. Tapetal origin of pollen coat substances in *Lilium*. *New Phytol.* 67, 779–86.
—— 1975. The physiology of the pollen grain surface. *Proc. R. Soc. Lond.* B 190, 275–99.

—— 1976. The adaptive significance of the exine. In: *The Evolutionary Significance of the Exine*, eds I. K. Ferguson and J. Muller. Linn. Soc. Symposium series, no. 1, pp. 27–38.

—— 1978. Recognition and response in the pollen-stigma interaction. *SEB Symposium*, no. 32, pp. 121–38.

—— 1979*a*. An interpretation of the hydrodynamics of pollen. *Am. J. Bot.* **66**, 737–43.

—— 1979*b*. Pollen walls as adaptive systems. *Ann. Miss. Bot. Gard.* (in Press).

—— and HESLOP-HARRISON, Y., 1970. Evaluation of pollen viability by enzymically induced fluorescence: intracellular hydrolysis of fluorescein diacetate. *Stain Technol.* **45**, 115–20.

—— —— 1979. Function of the Zwischenkörper in the germination of grass pollens. (In prep.)

—— —— and BARBER, J., 1975. The stigma surface in incompatibility responses. *Proc. R. Soc. Lond.* B **188**, 287–97.

—— —— KNOX, R. B. and HOWLETT, B., 1973. Pollen-wall proteins: 'gametophytic' and 'sporophytic' fractions in the pollen walls of Malvaceae. *Ann. Bot.* **37**, 403–12.

—— KNOX, R. B. and HESLOP-HARRISON, Y., 1974. Pollen-wall proteins: exine-held fractions associated with the incompatibility response in the Cruciferae. *Theor. Appl. Genet.* **44**, 133–7.

—— —— —— and MATTSSON, O., 1975. Pollen-wall proteins: emission and role in incompatibility responses. In *The Biology of the Male Gamete*, ed. J. G. Duckett and P. A. Racey. *Biol. J. Linn. Soc.* **7**, Suppl. 1, 189–202.

HESLOP-HARRISON, Y., 1977. The pollen–stigma interaction: pollen tube penetration in *Crocus. Ann. Bot.* **41**, 913–22.

HOEKSTRA, F. A. and BRUINSMA, J., 1978. Reduced independence of the male gametophyte in angiosperm evolution. *Ann. Bot.* **42**, 759–62.

HOSHIKAWA, K. 1960. Studies on pollen germination and pollen tube growth in relation to fertilisation in wheat. *Proc. Crop Sci. Soc. Japan* **28**, 333–6.

IDLE, D. B., 1966. The photography of ice formation in plant tissues. *Ann. Bot.* **38**, 199–206.

IWANAMI, Y., 1959. J. Yokohoma Municipal University C 34, 1 (cited in *Pollen, Biology, Biochemistry and Management*, eds R. G. Stanley and H. F. Linskens, pp. 307. Springer-Verlag, New York, Heidelberg, Berlin; original not seen).

IWANAMI, Y. and NAKAMURA, N., 1972. Storage in an organic solvent as a means for preserving viability of pollen grains. *Stain Technol.* **47**, 137–9.

JENSEN, W. A., 1962. *Botanical Histochemistry*, pp. 408. Freeman, San Francisco.

KESSLER, G., 1958. Zur Charakterisierung der Siebröhrenkallose. *Ber. Schweiz. Bot. Ges.* **68**, 5–43.

KIHARA, H. and HORI, T., 1966. The behaviour of nuclei in germination pollen grains of wheat, rice and maize. *Züchter* **36**, 145–50.

KNOWLTON, H. E., 1921. Studies in pollen, with special reference to longevity. *Cornell Univ. Agr. Exp. Stn Mem.* **52**, 745–93.

KNOX, R. B., 1971. Pollen-wall proteins: localisation, enzymic and antigenic activity during development in *Gladiolus. J. Cell Sci.* **9**, 209–37.

—— and HESLOP-HARRISON, J., 1970. Pollen-wall proteins: localisation and enzymic activity. *J. Cell Sci.* **6**, 1–27.

—— —— 1971(*a*). Pollen-wall proteins: electron microscopic localisation of acid phosphatase in the intine of *Crocus vernus. J. Cell Sci.* **8**, 727–33.

—— —— 1971(*b*). Pollen-wall proteins: the fate of intine-held antigens on the stigma in compatible and incompatible pollinations of *Phalaris tuberosa* L. *J. Cell Sci.* **9**, 239–51.

KROH, M. and LOEWUS, F., 1968. Biosynthesis of pectic substances in germinating pollen; labelling with myo-inositol-2^{14}C. *Science, N.Y.* **160**, 1352–4.

LABARCA, C. and LOEWUS, F., 1972. The nutritional role of pistil exudate in pollen tube wall formation in *Lilium longiflorum* L. I. Utilisation of injected stigmatic exudate. *Plant Physiol.* **50**, 7–14.

LANGE, W. and WOJCIECHOWSKA, B. 1976. The crossing of common wheat (*Triticum aestivum* L.) with cultivated rye (*Secale cereale* L.). I. Crossability, pollen grain germination and pollen tube growth. *Euphytica* **25**, 609–20.

LATO, M., BRUNELLI, B. and GIUFFINIA, B., 1969. Thin-layer chromatography of sugars on silica gel impregnated with sodium acetate, monosodium phosphate and disodium phosphate. *J. Chromat.* **39**, 407–16.

LEV, R. and SPICER, S. S., 1964. Specific staining of sulphate groups with alcian blue at low pH. *J. Histochem. Cytochem.* **12**, 309–10.

LOEWUS, F. and LABARCA, C., 1973. Pistil secretion product and pollen tube wall formation. In *Biogenesis of Plant Cell Wall Polysaccharides*, ed. F. Leowus, 503 pp. Academic Press, New York.

LINSKENS, H. F., 1967. Pollen. In *Handbuch der Pflanzenphysiologie*, ed. H. F. Linskens, vol. 18, pp. 368–406. Springer-Verlag, Berlin and New York.

—— and ESSER, K., 1957. Über eine spezifische Anfarbung der Pollenschläuche im Griffel und die zahl der Kallosepropfen nach Selbstung und Fremdung. *Naturwissenschaften* **33**, 16.

LUFT, J. H., 1964. Electron microscopy of cell extraneous coats as revealed by ruthenium red staining. *J. Cell Biol.* **23**, 55A.

LUXOVA, M., 1967. Fertilisation of barley (*Hordeum distichum* L.). *Biologia Pl.* **9**, 301–7.

MAEDA, H. and ISHIDA, N., 1976. Specificity of binding of hexapyranosyl polysaccharides with fluorescent brightener. *J. Biochem. (Tokyo)* **62**, 276–8.
MANGIN, L., 1889. Recherches sur le pollen. *Bull. Soc. Bot. Franc.* **33**, 512–7.
MAZIA, D., BREWER, P. A. and ALFERT, M., 1953. The cytochemical staining and measurement of protein with mercuric bromophenol blue. *Biol. Bull.* **104**, 527–40.
MEYER, H., 1971. Blütenbiologisches Untersuchungen bei Betreidekreuzung L. Das Wachstum von Roggen- und Weizenpollenschläuchin der Weisennarbe und die Entwicklung des Fruchtknotens bis 48 stunden nach der Bestaubung. *Archs Züchtungsforsch.* **1**, 171–83.
NAGATA, T. and TAKEBE, I., 1970. Cell wall regeneration and cell division in isolated tobacco mesophyll protoplasts. *Planta* **92**, 301–8.
NIETHAMMER, A., 1932. Die Pollenkeimung und chemisches Reizwirkungen in Zusammenhange mit der Mikrochemie des Kernes. *Biochem. Z.* 249–65.
NORTHCOTE, D. H., 1972. Chemistry of the plant cell wall. *A. Rev. Pl. Physiol.* **23**, 113–32.
PEARSE, A. G. E., 1972. *Histochemistry: Theoretical and Applied.* 3rd edn., vol. 2, 762–1518 pp. Churchill-Livingstone, London, Edinburgh.
PFAHLER, P. F. and LINSKENS, H. F., 1972. In vitro germination and pollen tube growth of maize (*Zea mays* L.) pollen. VII. Storage temperature and pollen source effects. *Theoret. App. Genet.* **42**, 136–40.
PFUNDT, M., 1910. Der Einflus der Luftfeuchtligkeit auf der Lebensdauer des Blütenstaubes. *Jahrb. Wiss. Bot.* **47**, 1–40.
PIHL, E., GUSTAFSON, G. T. and FALKMER, S., 1968. Ultrastructural demonstration of cartilage acid glycosaminglycans. *Histochem. J.* **1**, 26–35.
PODDUBNAYA-ARNOLDI, V. A., 1976. *Cytoembryology of Angiosperms. Principles and Perspectives*, 507 pp. Moscow (in Russian).
POPE, N. M., 1937. The course of the pollen tube in barley. *J. Am. Soc. Agron.* **38**, 432–40.
RANDOLPH, L. F., 1936. Developmental morphology of the caryopsis in maize. *J. Agric. Res.* **53**, 881–916.
ROGGEN, H. P. J. R. and STANLEY, R. G., 1969. Cell wall hydrolysing enzymes in wall formation as revealed by pollen tube extension. *Planta* **84**, 295–303.
ROLAND, F., 1971. Characterisation and extraction of the polysaccharides of the intine and the generative cell wall in the pollen grains of some of the Ranunculaceae. *Grana* **11**, 101–6.
ROSEN, W. G., 1968. Ultrastructure and physiology of pollen. *A. Rev. Pl. Physiol.* **19**, 435–62.
—— 1971. Pollen tube growth and fine-structure. In *Pollen, Development and Physiology*, ed. J. Heslop-Harrison, 338 pp. Butterworths Scientific Publications, London.
—— GALIK, S. R., DASHEK, W. V. and SIEGESMUND, K. A., 1964. Fine structure and cytochemistry of *Lilium* pollen tubes. *Am. J. Bot.* **51**, 60–71.
ROWLEY, J. R., 1960. The exine structure of 'cereal' and 'wild' type grass pollen. *Grana Pal.* **2**, 9–15.
—— 1964. Formation of the pore in pollen of *Poa annua*. In *Pollen Physiology and Fertilisation*, ed. H. F. Linskens, 257 pp. North Holland, Amsterdam.
—— MÜHLETHALER, K. and FREY-WYSSLING, A., 1959. A route for the transfer of materials through the pollen grain wall. *J. Biophys. Biochem. Cytol.* **6**, 537–8.
SARTORIS, G. B., 1942. Longevity of sugar cane and corn pollen – a method for long-distance shipment of sugar cane pollen by airplane. *Am. J. Bot.* **29**, 395–403.
SASSEN, M. M. A., 1964. Fine structure of *Petunia* pollen grain and pollen tube. *Acta Bot. Neerl.* **13**, 175–81.
SHAW, G. and YEADON, A., 1966. Chemical studies on the composition of some pollen and spore membranes. *J. Chem. Soc.* C pp. 16–22.
SHIVANNA, K. R., HESLOP-HARRISON, Y. and HESLOP-HARRISON, J. S., 1978. Inhibition of the pollen tube in the self-incompatibility response of grasses. *Incomp. Newslett.* **10**, 5–7.
SIMON, E. W., 1974. Phospholipids and plant membrane permeability. *New Phytol.* **73**, 337–420.
—— 1978. Plant membranes under dry conditions. *Pectic. Sci.* **9**, 169–72.
SITTE, P., 1953. Untersuchungen zur submikroskopischen Morphologies der Pollen and Sporenmembranen. *Mikroskopie* **8**, 290–9.
SKVARLA, J. J. and LARSON, D. A., 1966. Fine-structural studies of *Zea mays* pollen. I. Cell membranes and exine ontogeny. *Am. J. Bot.* **53**, 1112–25.
SOUTHWORTH, D., 1974. Solubility of pollen exines. *Am. J. Bot.* **61**, 36–44.
STANLEY, R. G. and LINSKENS, H. F., 1974. *Pollen: Biology, Chemistry and Management*, 307 pp. Springer-Verlag, New York, Heidelberg, Berlin.
TAKEUCHI, Y. and KOMAMINE, A., 1978. Composition of the cell wall formed by protoplasts isolated from cell suspension cultures of *Vinca rosea*. *Planta* **140**, 227–32.
VAN DER WOUDE, W. J. and MORRÉ, D. J., 1967. Endoplasmic reticulum–dictyosome–secretory vesicle associations in pollen tubes of *Lilium longiflorum* Thunb. *Proc. Indiana Acad. Sci.* **77**, 164–70.
—— —— and BRACKER, C. E., 1971. Isolation and characterisation of secretory vesicles in germinated pollen of *Lilium longiflorum*. *J. Cell Sci.* **8**, 331–51.
VASIL, I. K., 1960. Pollen germination in some Gramineae: *Pennisetum typhoideum*. *Nature, Lond.* **187**, 1134–5.
—— 1964. Effect of boron on pollen germination and pollen tube growth. In *Pollen Physiology and Fertilisation*, ed. H. F. Linskens, 257 pp. North Holland Publishing Co., Amsterdam.

VINSON, C. G., 1927. Some nitrogenous constituents of corn pollen. *J. Agric. Res.* **35**, 261–78.
VITHANAGE, H. I. M. V. and HESLOP-HARRISON, J., 1979. The pollen-stigma interaction: fate of fluorescent-labelled pollen wall proteins on the stigma surface in rye (*Secale cereale*). *Ann. Bot.* **43**, 113–4.
—— and KNOX, R. B., 1977. Development and cytochemistry of the stigma surface and response to self and foreign pollination in *Helianthus annuus*. *Phytomorphology* **27**, 168–79.
WATANABE, K., 1955. Studies on the germination of grass pollen. I. Liquid exudation of the pollen on the stigma before germination. *Bot. Mag. Tokyo* **68**, 40–4.
—— 1961. Studies on the germination of grass pollen. II. Germination capacity of pollen in relation to the maturity of pollen and stigma. *Bot. Mag. Tokyo* **74**, 131–7.
ZEEVEN, A. C. and VAN HEEMERT, C., 1970. Germination of weed rye (*Secale segetale* L.) on wheat (*Triticum aestivum* L.) stigmas and the growth of the pollen tubes. *Euphytica* **19**, 175–9.

EXPLANATION OF PLATES

Abbreviations used: E, exine; I, intine; Z, Zwischenkörper; Zl, Z-layer; B, baculum; S, sexine; N, nexine; A, aperture; MW, microfibrillar wall; C, callosic wall; M, mitochondrion.

PLATE 1

Germination apertures of grass pollens.

A. Scanning electron micrograph of the aperture and adjacent exine of *Poa pratensis*. The aperture is surrounded by a raised lip of the exine, and the central operculum is borne on a thin sporopollenin lamella, overlying the Zwischenkörper. In this instance additional sculptured fragments are scattered over the lamellae. ×c. 3600.

B. Transmission electron micrograph of a sectioned aperture of *Sorghum bicolor*. O, operculum. The Zwischenkörper shows continuity with the Z-layer (Zl). The intine, seen with cytoplasmic inclusions, is still in the course of thickening. This micrograph, which shows the Zwischenkörper and Z-layer exceptionally clearly, is reproduced through the courtesy of Drs J. E. Christensen and H. T. Horner. ×c. 25000.

PLATE 2

Transmission electron micrographs of hydrating pollen of *Secale cereale* fixed in OsO_4 vapour.

A. Glancing section cutting the sexine and nexine but not the surface cytoplasm of the vegetative cell. The surface exudate (El) has been preserved *in situ* with little diffusion. Bacula are seen in section. ×c. 11000.

B. As A, showing detail of sexine and distribution of micropores. ×c. 28000.

C. As B, nexine. The micropores are less frequent than in the sexine. ×c. 28000.

PLATE 3

Pollen of *Secale cereale*.

A. Fluorescence optical micrograph of 0·5 μm section of an ungerminated grain, auramine O staining. The sporopollenin of the exine and operculum fluoresces with this stain, and also lipid inclusions in the cytoplasm. The starch grains appear in dark contrast. ×c. 700.

B. As A, germinating grain. The Zwischenkörper has gelatinized, and may be seen as a prominence at the aperture; the starch grains are now oriented towards the aperture. G, generative nuclei. ×c. 700.

C. Transmission electron micrograph of an oblique section of the wall of a mature hydrating grain fixed in OsO_4 vapour. The intine inclusions are dilated, and many appear tubular in section. ×c. 43000.

D. Sexine and bacula in radial section. The sexine shows the dark outer surface-layer. ×c. 78000.

PLATE 4

Pollen of *Secale cereale*.

A. Fluorescence light micrograph of a 0·5 μm section of an ungerminated grain, calcofluor white staining. The intine fluoresces strongly with this staining; it shows the characteristic configuration at the aperture site. The exine is visible because of the slight autofluorescence of sporopollenin. The operculum is seen to lie on a sporopollenin lamella attached to the rim of the surrounding exine annulus. The Zwischenkörper lies between this and the intine. ×c. 1200.

B. As A, auramine O staining. The intine is not visible, but the exine, operculum and intervening lamella are strongly fluorescent. ×c. 1200.

C. Aperture and adjacent wall, alcian blue staining. The Zwischenkörper is heavily stained, but not the overlying operculum or underlying intine. ×c. 1000.

D. Apertural region of an unfixed freeze-sectioned grain; acid phosphatase localization, α-naphthyl acid

phosphate substrate, hexazotized pararosanilin coupling agent. Activity is strongly localized to the intine, but some reaction product is also associated with the Zwischenkörper, probably through diffusion of the enzyme during preparation. $\times c.$ 1200.
E. As C, alcian blue staining, surface view of isolated wall showing Zwischenkörper region. The operculum has been displaced. $\times c.$ 800.
F. As E, operculum retained; the heavily stained Zwischenkörper is visible through the operculum and the surrounding lamella. $\times c.$ 1000.

PLATE 5

Pollen of *Secale cereale*, preparation of intine 'ghosts'.
A. Intermediate stage in the dissolution of the exine, auramine O staining, fluorescence light micrograph. The sexine has loosened and separated from the nexine. $\times c.$ 520.
B. Intine ghosts after removal of the exine and alkaline digestion of the cell contents. Calcofluor white staining, fluorescence micrograph. $\times c.$ 400.
C. Birefringence of the apertural region of an intine ghost. $\times c.$ 900.
D. Pollen after $c.$ 20 min exposure to digestion by cellulase (*Aspergillus niger* source) after removal of the exine, but without extraction of cell content. The intines are undergoing dissolution. Calcofluor white staining, fluorescence micrograph. $\times c.$ 200.
E. Exine-less pollen after exposure to digestion by helicase for $c.$ 2 h. The intines are entirely lost, and the protoplasts are coalescing. Calcofluor white staining, fluorescence micrograph. $\times c.$ 310.

PLATE 6

Starch in the vegetative cell of *Secale cereale*.
A. Extruded protoplast from an unfixed vegetative cell. I–KI staining, light micrograph. $\times c.$ 750.
B. Starch in pollen macerate, ensheathed in residual plastid protein. Coomassie blue staining, light micrograph. $\times c.$ 900.
C. Pollen macerate, PAS staining, light micrograph. The P-particles are PAS-reactive, and form the granular background to the starch; the light inclusions are lipid globuli. $\times c.$ 1500.
D. Late germination, PAS staining, showing eroding starch. Light micrograph of 0·5 μm section. $\times c.$ 700.
E. Electron micrograph of an amyloplast in ungerminated pollen, uranyl acetate-KMnO$_4$ post-staining. $\times c.$ 45000.
F. Smear from the vegetative cell of an ungerminated pollen grain, showing starch in a proteinaceous matrix. Coomassie blue staining, light micrograph. $\times c.$ 800.

PLATE 7

Electron micrographs of ungerminated pollen of *Secale cereale*; glutaraldehyde-OsO$_4$ fixation, uranyl acetate-KMnO$_4$ post-staining.
A. Periphery of vegetative cell, with irregular lipid inclusions at the surface of the protoplast. In this ungerminated state the intine inclusions are poorly defined. $\times c.$ 22000.
B. As A, detail, showing packed P-particles and intervening lipid inclusions. $\times c.$ 52000.
C. As A, detail, showing mitochondria compressed between P-particles. The mitochondria have poorly defined membranes. $\times c.$ 50000.

PLATE 8

Light micrographs of pollen of *Secale cereale* (A, B and C) and *Phleum pratense* (D, E and F); hydration, germination and protein emission.
A. Pollen after 2 h natural drying at 23 °C and 50 per cent relative humidity. $\times c.$ 180.
B. As A, after 1 min rehydration in 0·5 M sucrose. $\times c.$ 180.
C. Nine frames from a time-lapse sequence showing hydration, exudation, resorption and germination of a single pollen grain on the stigma. The first frame shows the grain $c.$ 20 s after manual application; time intervals of 30 s thereafter. Germination in this instance was $c.$ 2 min 20 s after contact. The meniscus (Me) formed in 20–25 s, and this was followed rapidly by the first (apertural) exudation (Ex). T, Tube tip. The interrupted line in Frame 8 shows the approximate plane of the section in Plate 17B. $\times c.$ 235.
D, E and F. Protein emission from pollen in Coomassie blue stain-fixing medium. D, Early emission through the exine; E and F later stages of emission, mainly from the aperture. $\times c.$ 800.

Plate 9

Electron micrographs of germinating pollen of *Secale cereale*. Glutaraldehyde-OsO$_4$ fixation, uranyl acetate-KMnO$_4$ post-staining.

A. Early vacuolation. The lipids at the cell surface are dispersing, the P-particles are separating and the mitochondria are regaining a normal structure. × c. 25000.
B. Later germination stage. The mitochondria are now well defined, the tonoplast and membranes of the endoplasmic reticulum are resolved, and dictyosomes (D) are present, although not abundant. × c. 23500.

Plate 10

Light micrographs of germination and early tube growth in *Secale cereale*.

A. Gelation and dispersal of the Zwischenkörper in 0·6 M sucrose in dilute indian ink. The gel emerging from the aperture excludes the carbon particles. × c. 370.
B. As A; the gel has now accumulated a cap of carbon particles. × c. 500.
C. Early establishment of tip growth in the emerging tube. Organelles are already excluded from the clear apical zone of the tube, which is populated mainly with P-particles. The tube tip is secreting proteins into the medium, revealed by the darkly stained cap, which extends back only to the limit of the clear zone of the tube cytoplasm. × c. 1150.
D. As C, later stage. The clear tip zone is now c. 5 µm, and this length is maintained throughout further growth of the tube. Protein secretion is mainly from the flanks, but it extends back to, and overlaps somewhat with, the zone where the inner callose wall layer is being laid down. Coomassie blue stain-fixing medium after germination on Medium A. × c. 1150.
E. Actively growing pollen tube, phase contrast with electronic flash illumination, showing tip zonation. × c. 1100.
F. Actively growing tube, showing the secretion of proteins from the tip and sub-adjacent zone. Coomassie blue stain-fixing medium after germination on Medium A. × c. 900.

Plate 11

Fluorescence micrographs of intine ghosts of *Secale cereale*; calcofluor white staining.

A. Surface, and B, profile of ghosts from ungerminated grains. × c. 1100.
C. Surface views of ghosts extracted from germinating grains 90 s after the beginning of hydration. The weakening of the intine in register with the Zwischenkörper has permitted the first extrusion of the tube. × c. 1200.

Plate 12

Light micrographs of pollen of *Secale cereale*.

A. Intine ghost from germinating grain c. 150 s after the beginning of hydration, showing the emerging apertural intine in profile. Fluorescence micrograph, calcofluor white staining. × c. 1300.
B. As A, later stage. The tube is now growing actively; the tip is off to the right, and the micrograph shows the continuity of the wall with the intine. Fluorescence micrograph, calcofluor white staining. × c. 1300.
C. Intact germinating pollen, calcofluor white staining. Stage corresponding to A. The gelation and emergence of the Zwischenkörper has disrupted the overlying sporopollenin lamellae and lifted the operculum (O), which in this instance remains attached to the gel, but separated by it from the tube tip proper. The exine is visible through the autofluorescence of sporopollenin. × c. 830.
D. Pseudo-germination of an inviable grain. The Zwischenkörper has dispersed on hydration, and the intine has been forced out in a balloon-like hernia without the development of a normal tube tip. Fluorescence micrograph, calcofluor white staining. × c. 800.
E. Light micrograph of a 0·5 µm transverse section of two pollen tubes in contact with the stigma, PAS staining. The stratification of the tube wall may be distinguished, the outer layer PAS-reactive, and the inner, callose, layer unstained. × c. 1400.

Plate 13

Electron micrographs of early pollen tube growth in *Gaudinia fragilis*. Glutaraldehyde-OsO$_4$ fixation, uranyl acetate and lead citrate staining.

A. Oblique section of the emerging tube near the tip. The wall has a random microfibrillar structure, and at this level lacks an inner callosic lining. × c. 30000.
B. Tube wall and adjacent cytoplasm behind the growing tip. The microfibrillar part of the wall is sharply

delimited from the inner, homogeneous, callosic layer. The cytoplasm is crowded with P-particles and lipid bodies. ×c. 26000.

C. Glancing section of the tube wall in the immediate sub-apical zone. The inner callosic layer shows partly incorporated P-particles. Single membrane-bounded vesicles of smaller size are present at the interface of the cytoplasm and the wall. ×c. 30000.

D. Section of tube wall in the same zone as C, showing numerous single membrane-bounded vesicles near the callosic layer. ×c. 32000.

Plate 14

Electron micrographs of germinating pollen of *Secale cereale*. Glutaraldehyde-OsO$_4$ fixation, uranyl acetate-KMnO$_4$ post-staining.

A. Eroding starch with adjacent bodies with microfibrillar content similar to that of the P-particles at this stage of germination. ×c. 51000.

B. As A, with membrane profiles. ×c. 45000.

C. Lipid globuli and contiguous membrane-bounded bodies with microfibrillar content similar to that of the P-particles. ×c. 80000.

Plate 15

Electron micrographs of germinating pollen of *Secale cereale*. Glutaraldehyde-OsO$_4$ fixation, uranyl acetate and lead citrate post-staining.

A. Aperture with withdrawing protoplast. Intine ghosts prepared from pollen at this stage have the appearance seen in Plate 12B, since the removal of the exine eliminates the constriction in the shaft of the aperture. The tube wall is seen to be continuous with the intine, which is undergoing dissolution. Lipid globuli are leaving the grain, and also large microfibrillar bodies, the product of the coalescence of P-particles, or derived directly from the reserves of the grain, including the intine. ×c. 26000.

B. Detail of large microfibrillar body, with content similar to that of the P-particles, in the protoplast of the vegetative cell during the final withdrawal from the grain. These bodies are moderately PAS-reactive, and stain with calcofluor white (cf. Plate 16A). ×c. 70000.

C. Detail of A, showing association of the tube base with the residual intine and the microfibril orientation in the wall. ×c. 60000.

Plate 16

Electron micrographs of late germination stages in pollen of *Secale cereale*. Glutaraldehyde-OsO$_4$ fixation, uranyl acetate and lead citrate post-staining.

A. Dispersing intine, slightly earlier stage than in Plate 15A. The microfibrillar bodies derived from the intine have a content identical in appearance with that of the P-particles at this stage, and can be traced throughout the protoplast of the tube (cf. Plate 15B). ×c. 28000.

B. Exine of empty grain after complete withdrawal of the cytoplasm. The intine remains as a loose mesh of fibrils. ×c. 73000.

C. Residual membranes and vesicles in an empty grain after withdrawal of the cytoplasm. In this example essentially no intine polysaccharide remains. ×c. 20000.

Plate 17

Electron micrographs of pollen contents and tubes. Glutaraldehyde-OsO$_4$ fixation, uranyl acetate and lead citrate post-staining. A, B. *Secale cereale*; C, *Gaudinia fragilis*.

A. P-particles released from a disrupted ungerminated grain showing the heterogeneity of internal structure arising during hydration. ×c. 72000.

B. Germinating grain, profile showing part of the tube growing over the grain surface, the plane of section corresponding approximately to the line indicated on the 8th frame of the sequence of Plate 8c. The tube wall shows inner callosic and outer microfibrillar layers. Me, Fixed residue of the meniscus between the tube and the grain surface. The tube is packed with P-particles, and many remain in the grain. ×c. 11500.

C. Older part of a pollen tube, after penetration of the stigma. The tube is growing underneath the cuticle (Cu) of the stigma which is in direct contact with the microfibrillar part of the tube wall, itself clearly demarcated from the inner callosic lining at this time. The section is some 200 μm behind the growing tip. ×c. 19000.

PLATE 18

Light micrographs of pollen of Gramineae.

A. Release of pectinase from hydrating pollen of *Secale cereale*, dark field illumination. The dark areas mark the sites from which the pectin of the substrate film has been digested by the pollen enzyme, the grains having moved during the subsequent processing of the film. ×c. 190.

B. Ca^{2+}-blocked germination of pollen of *Hordeum bulbosum*, viewed after 40 min. Two grains are fully dilated and have no remaining starch content. The undilated grain is inviable, and retains starch content. ×c. 260.

C. Frames from a time-lapse sequence of a pollen grain of *Secale cereale*, Ca^{2+}-blocked germination. The grains are fully dilated after 60 min. The rate of cytoplasmic movement can be judged from the change of position of the cytoplasmic strands. Successive frames may be viewed as stereo pairs to enhance the differences. Electronic flash illumination, 7 s intervals. ×c. 320.

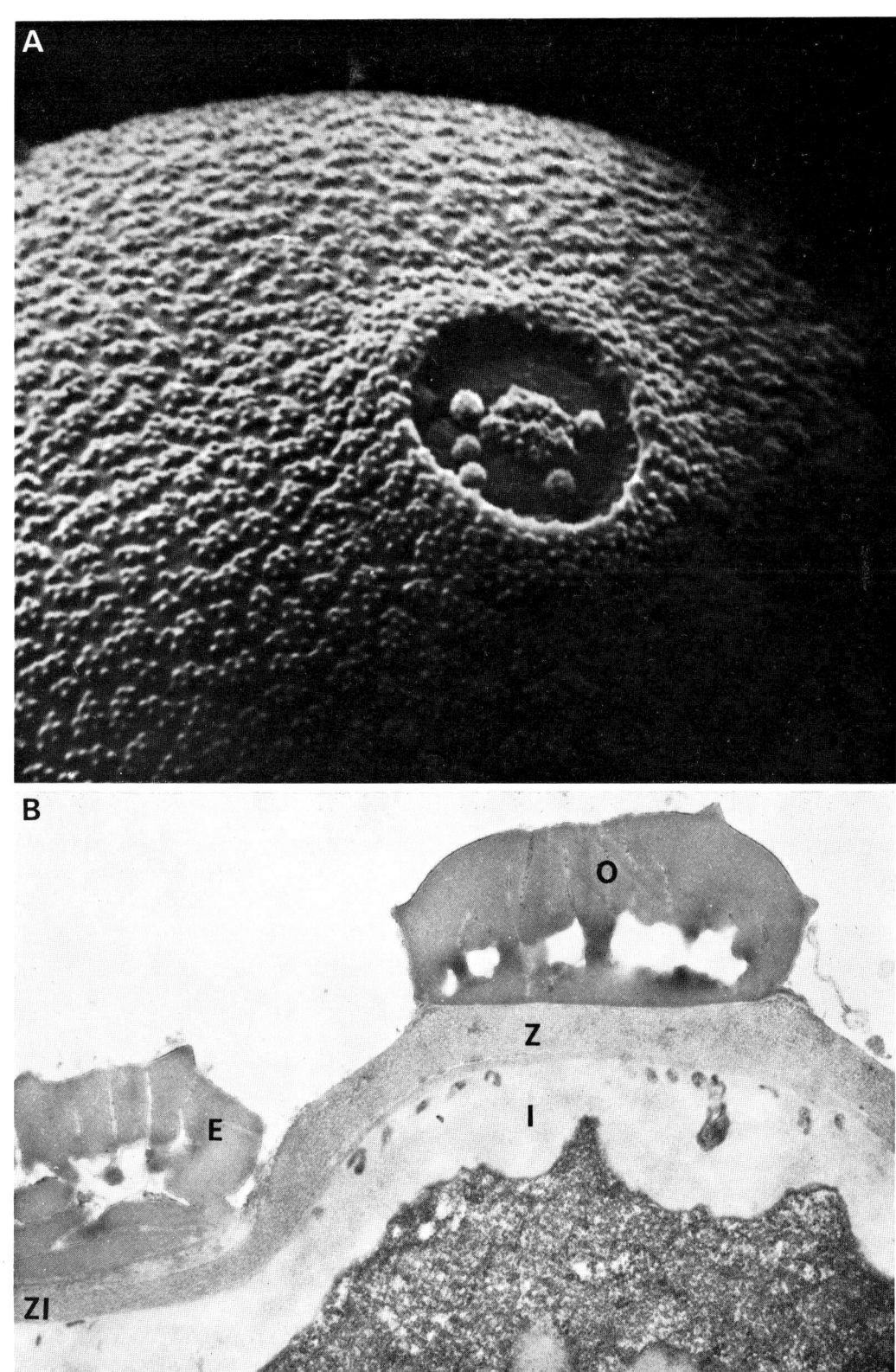

PLATE 1

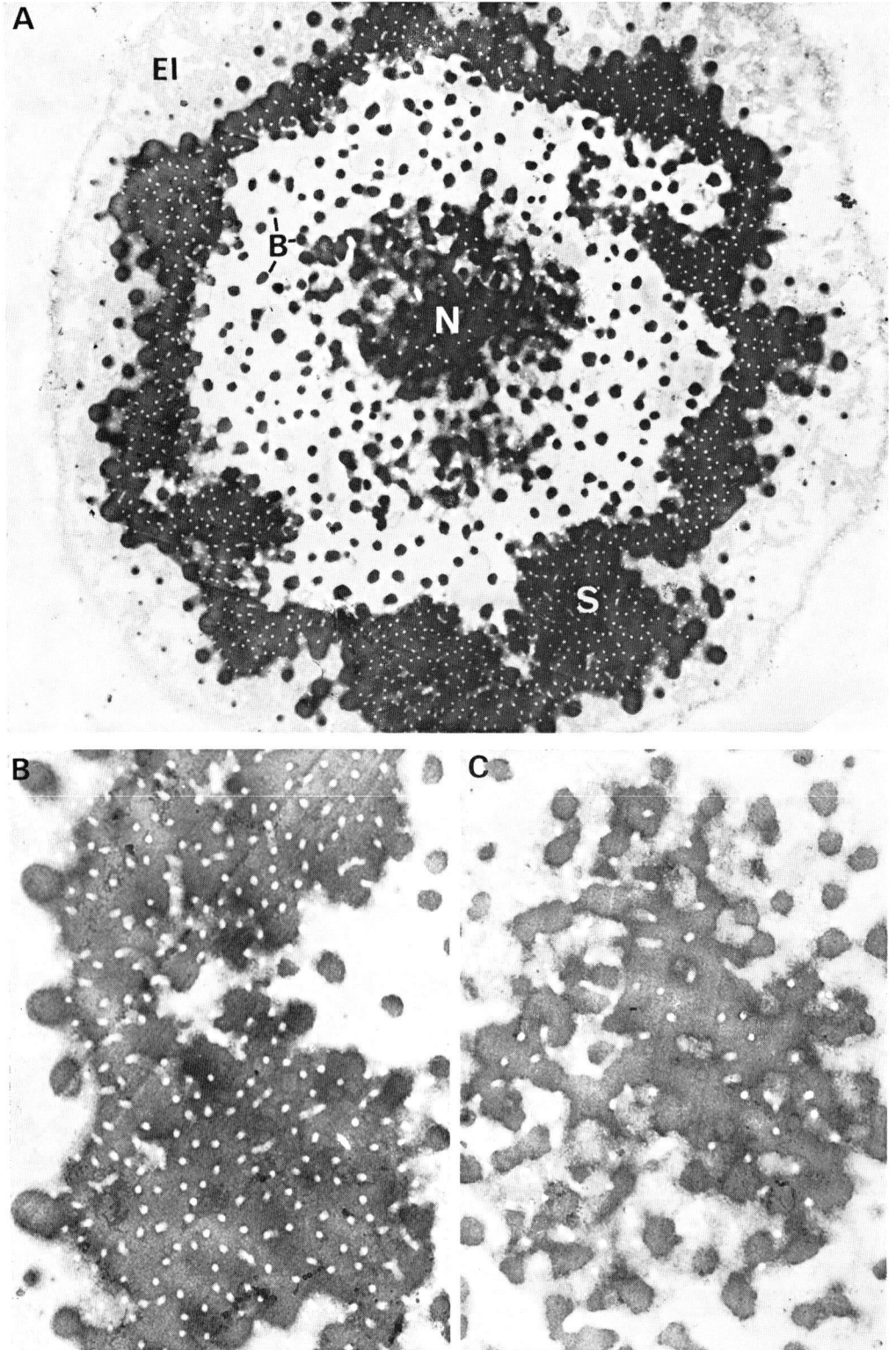

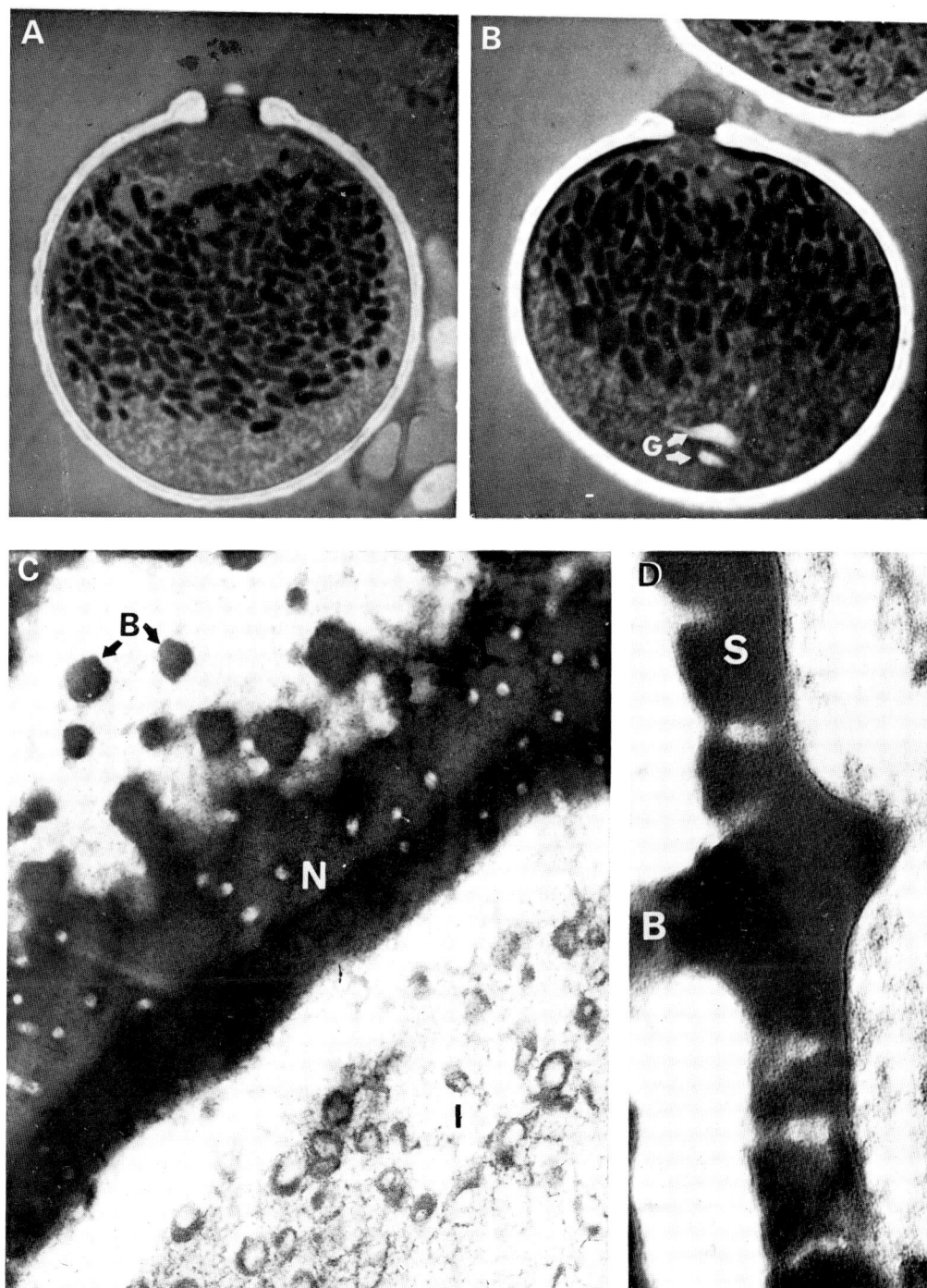

PLATE 3

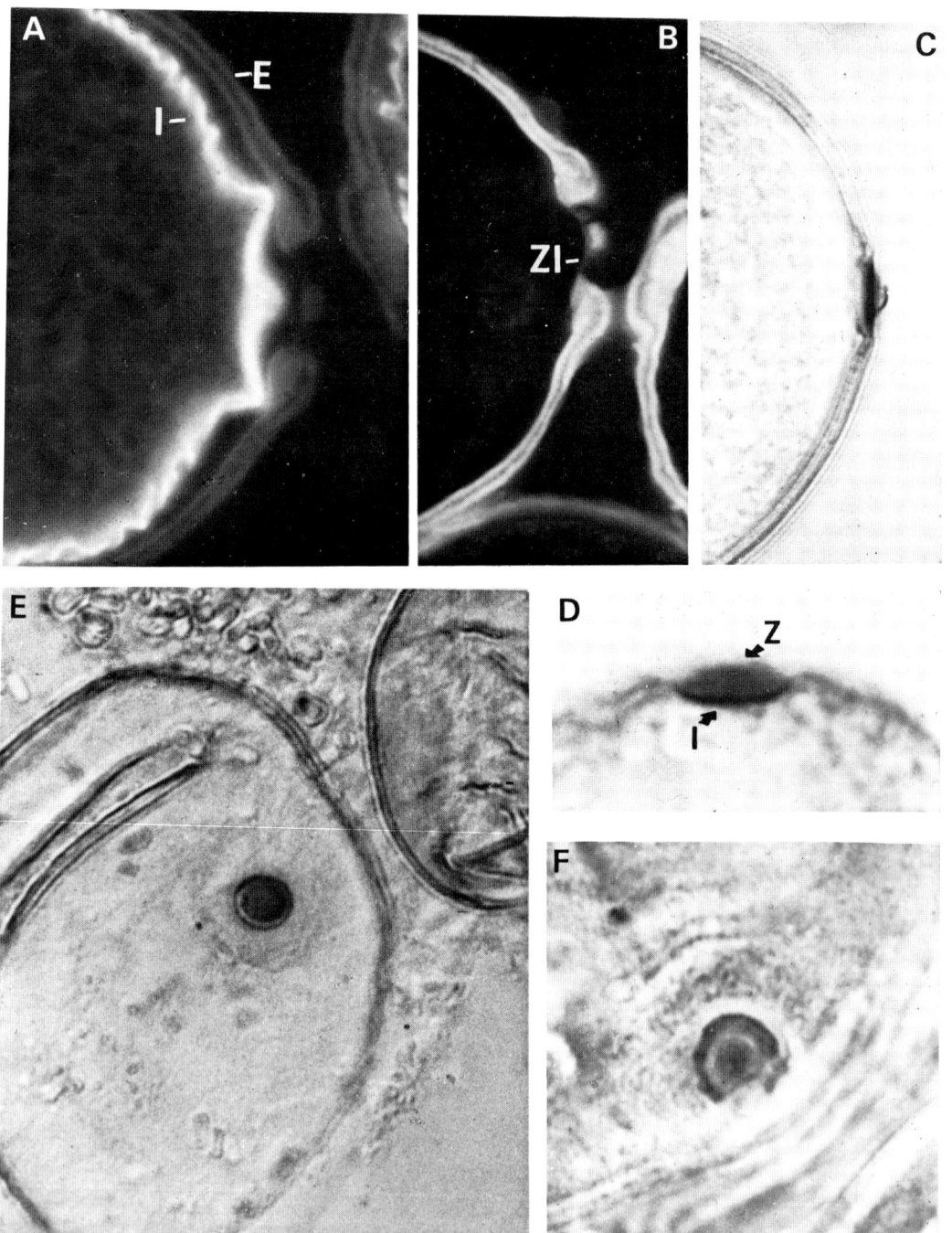

PLATE 4

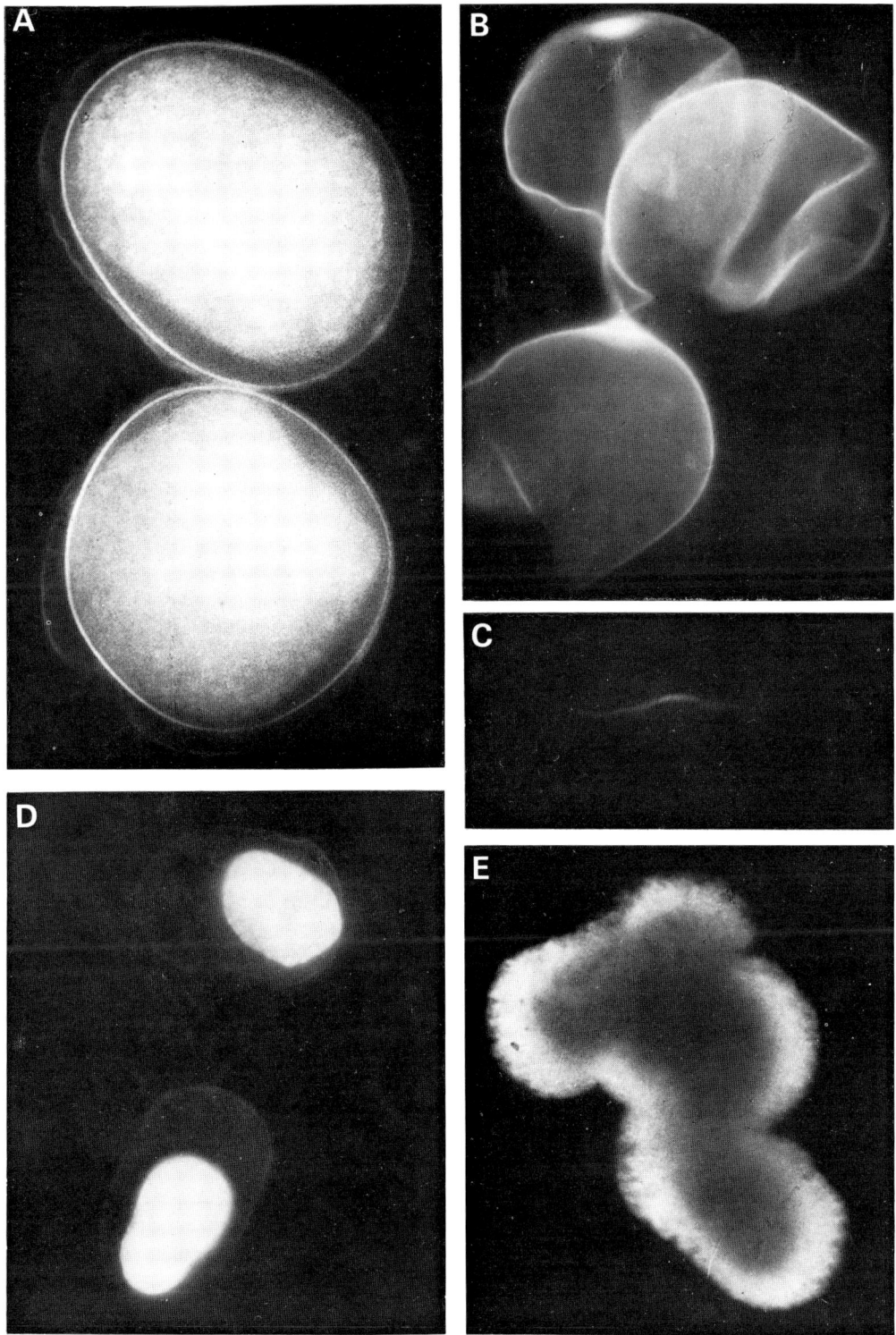

PLATE 5

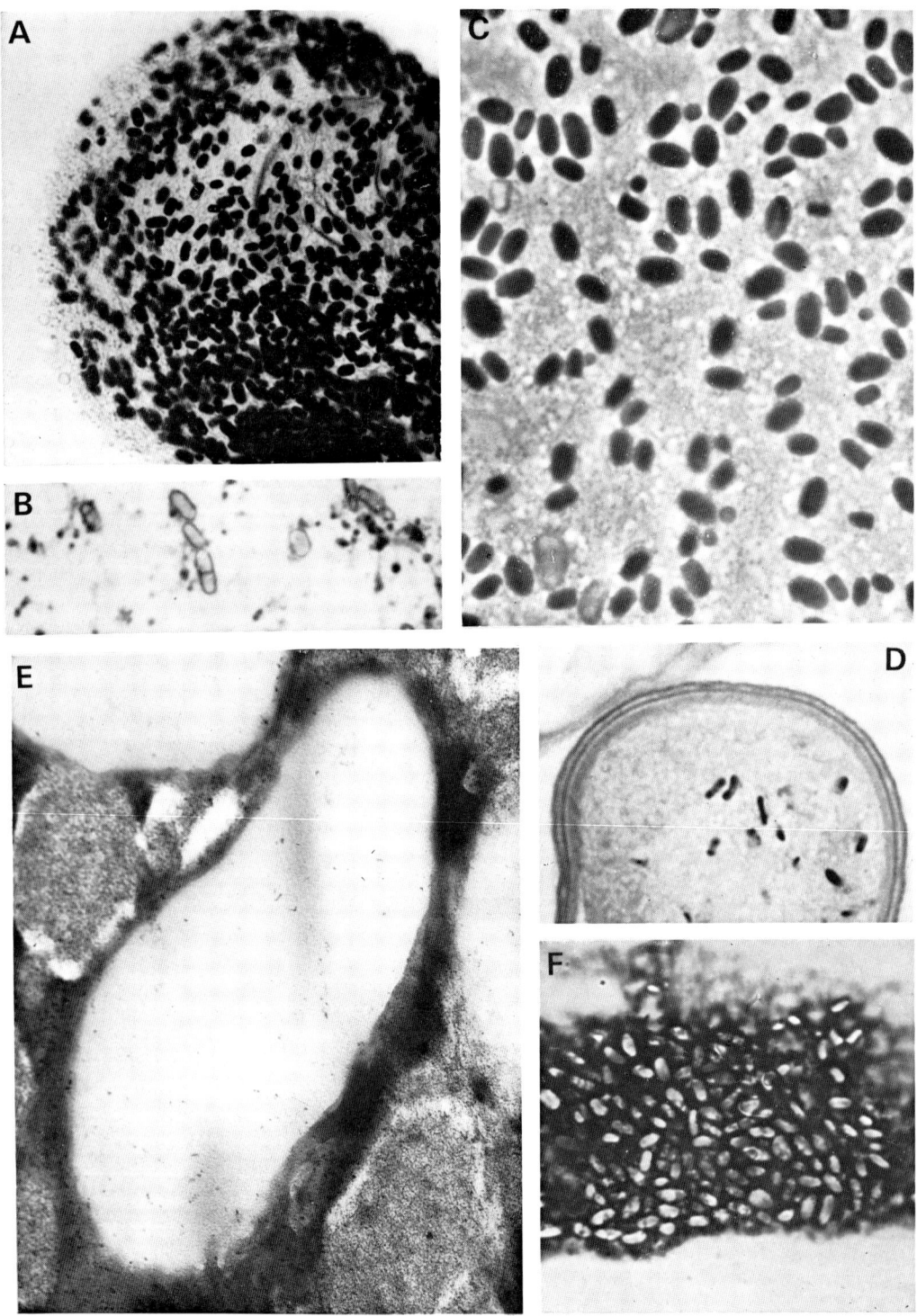

PLATE 6

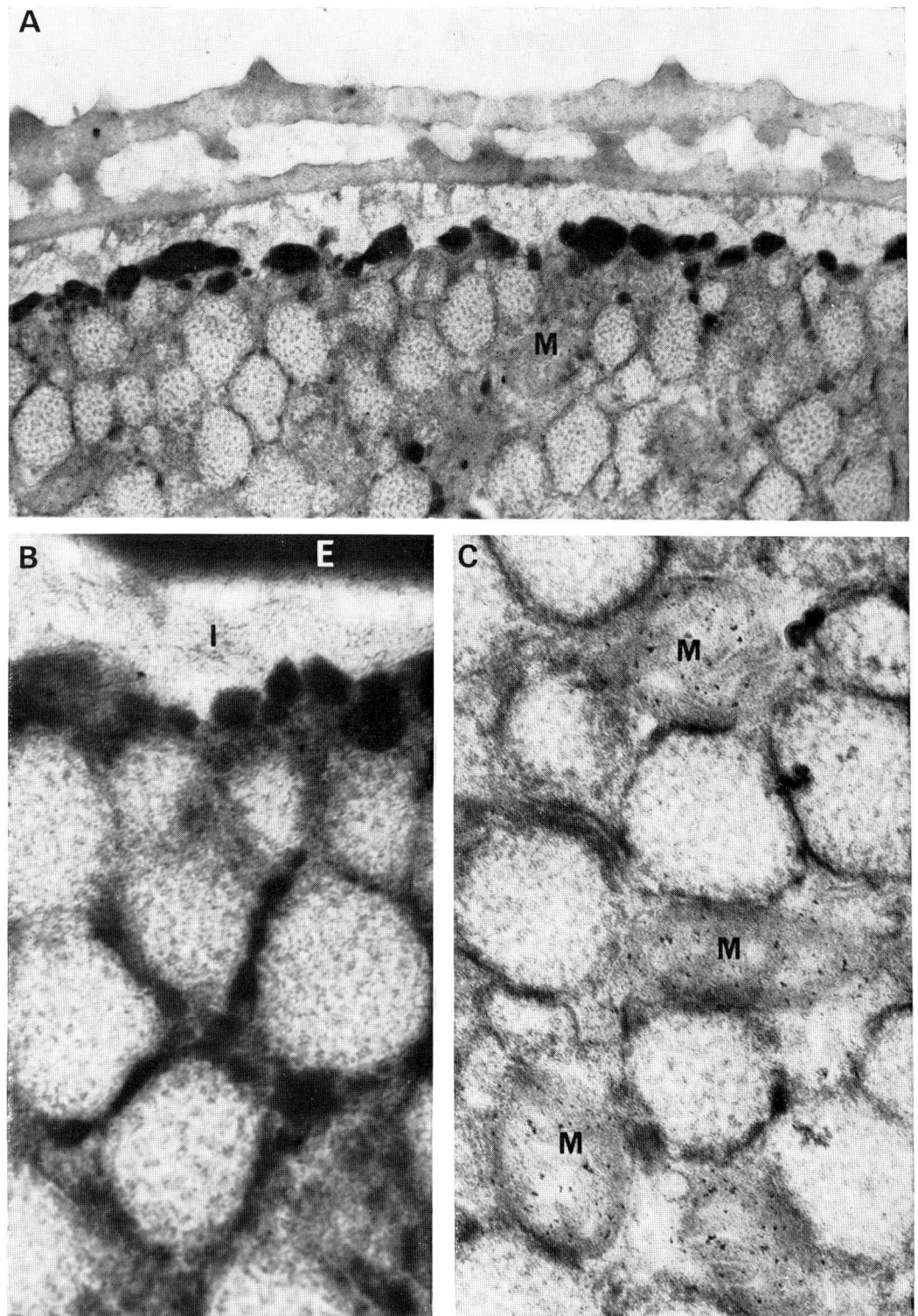

PLATE 7

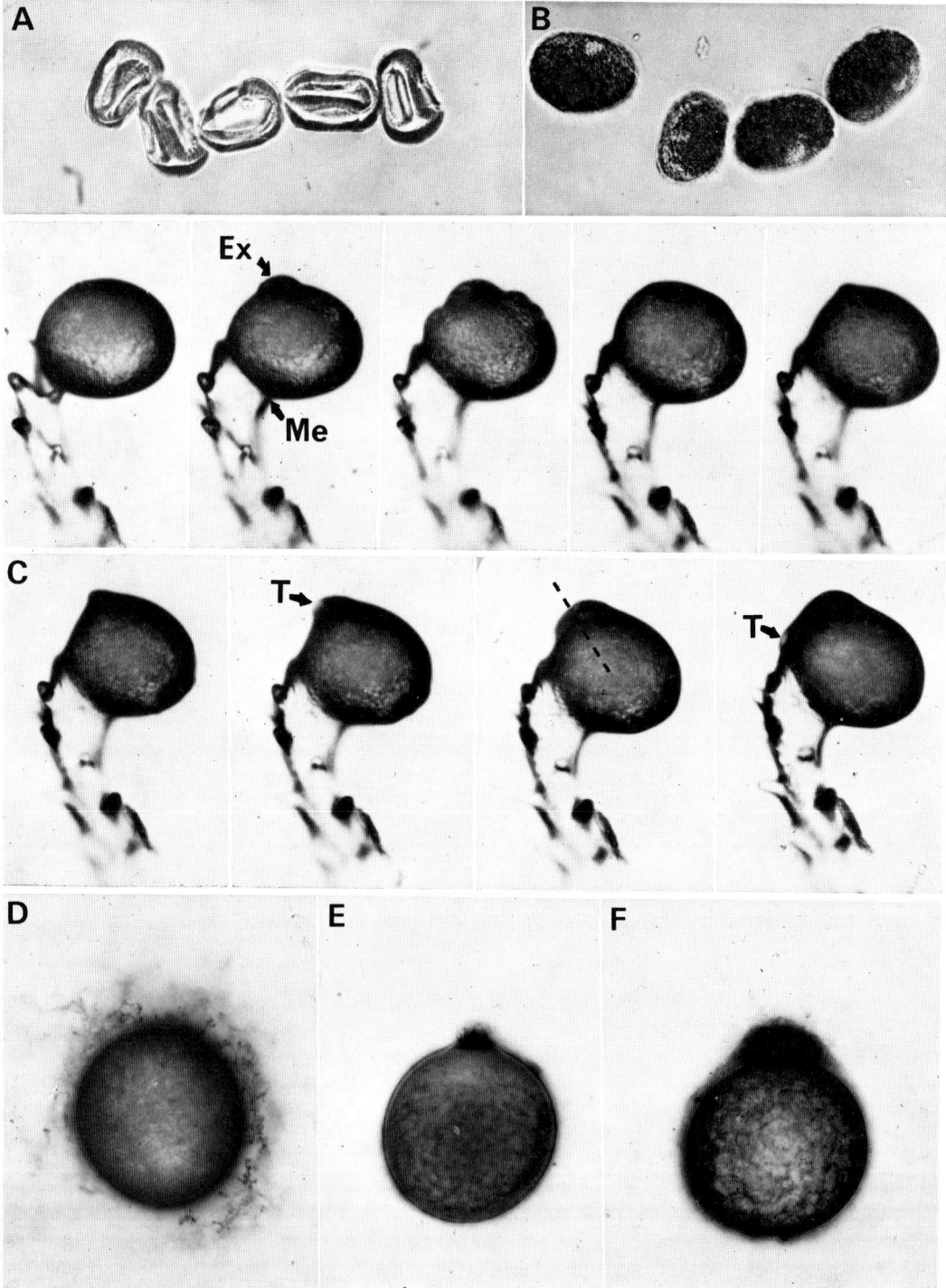

PLATE 8

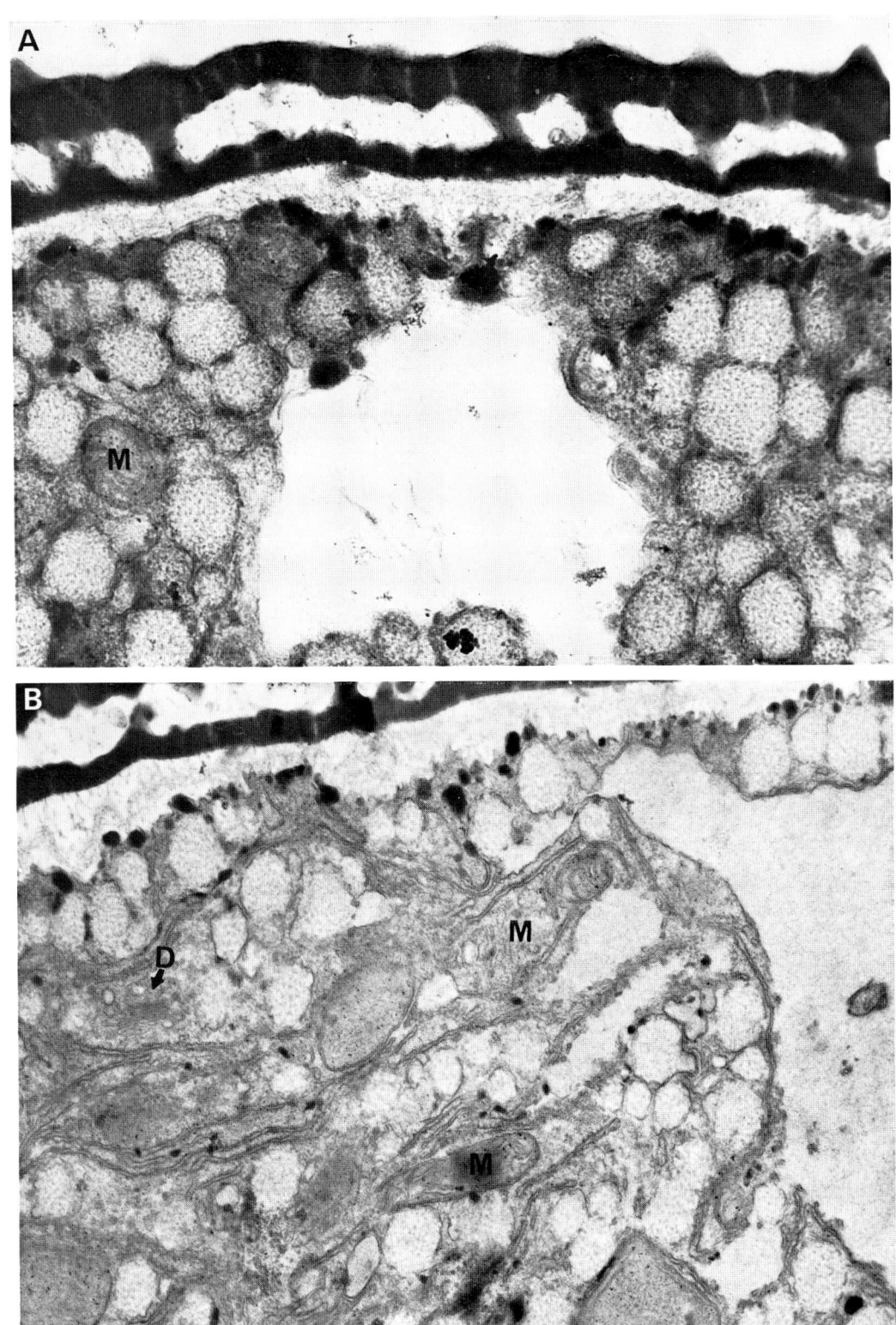

PLATE 9

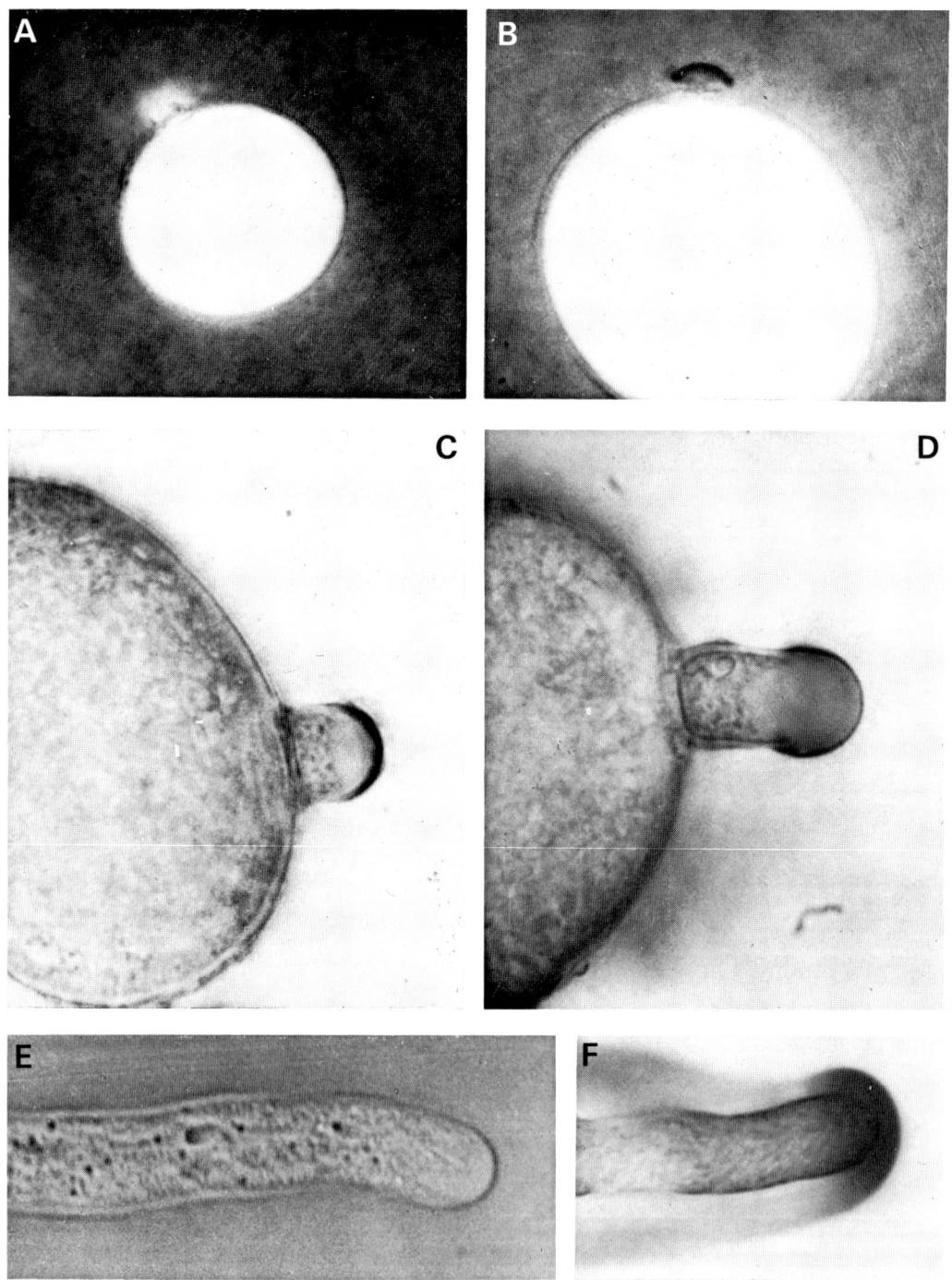

PLATE 10

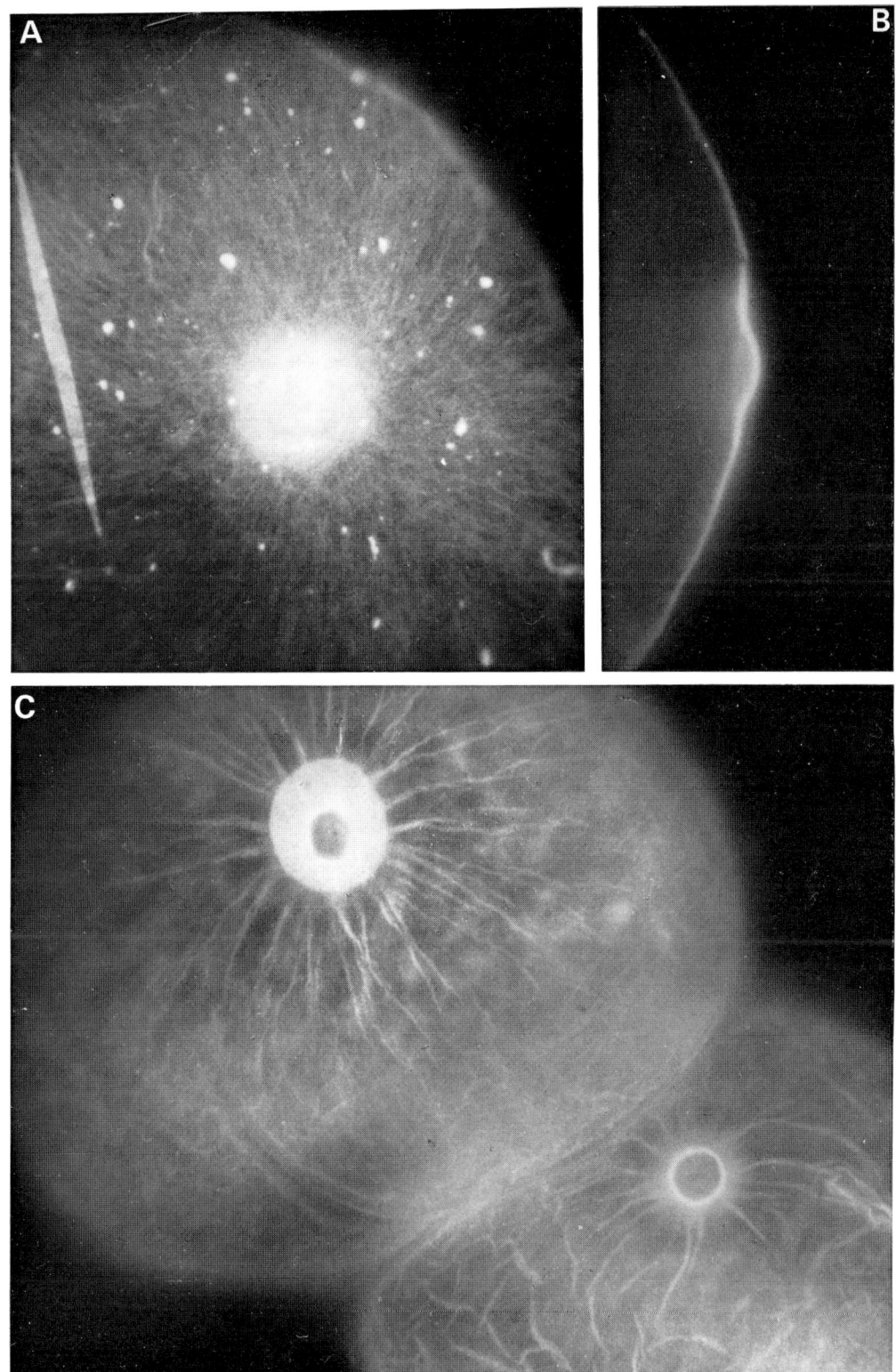

PLATE 11

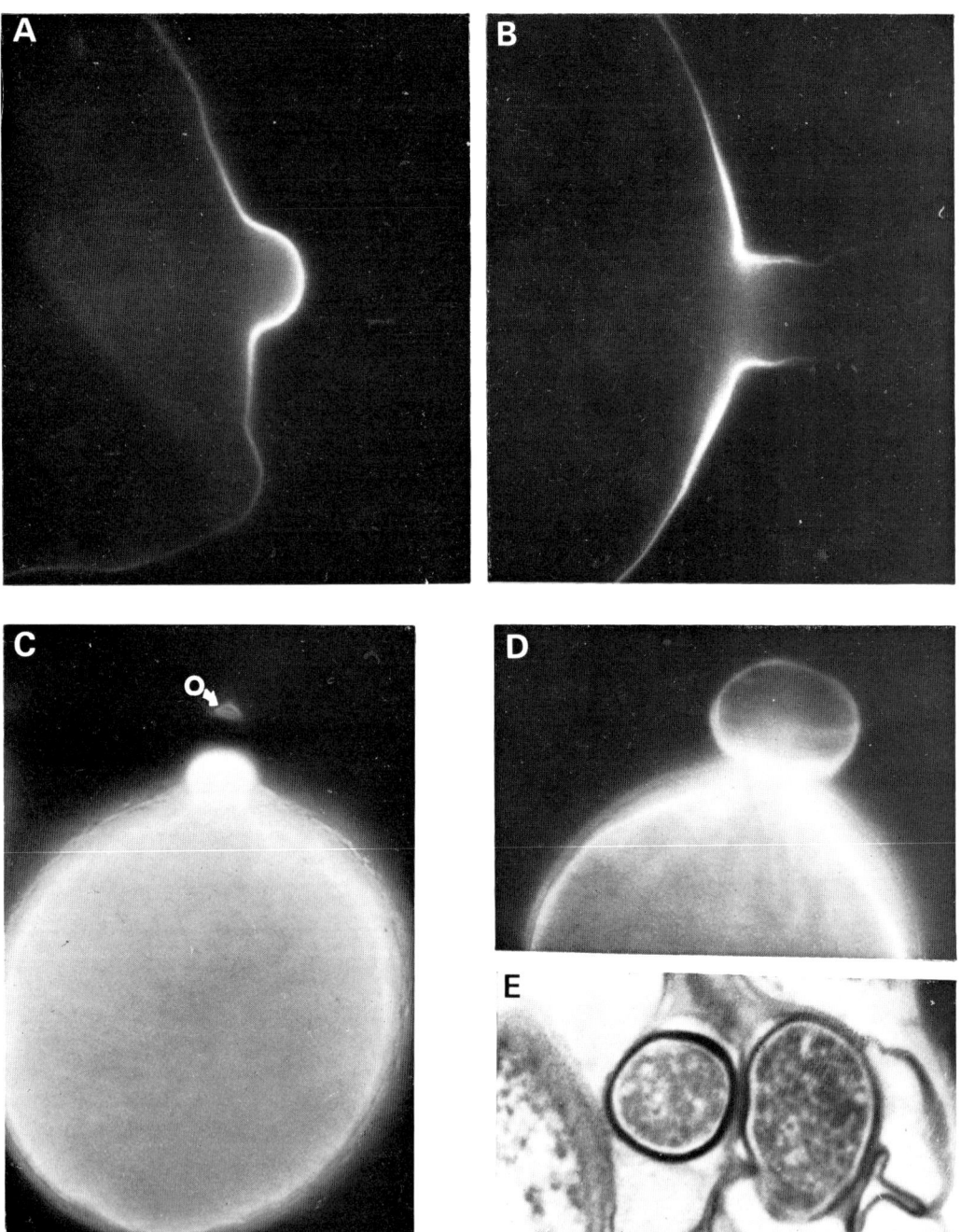

PLATE 12

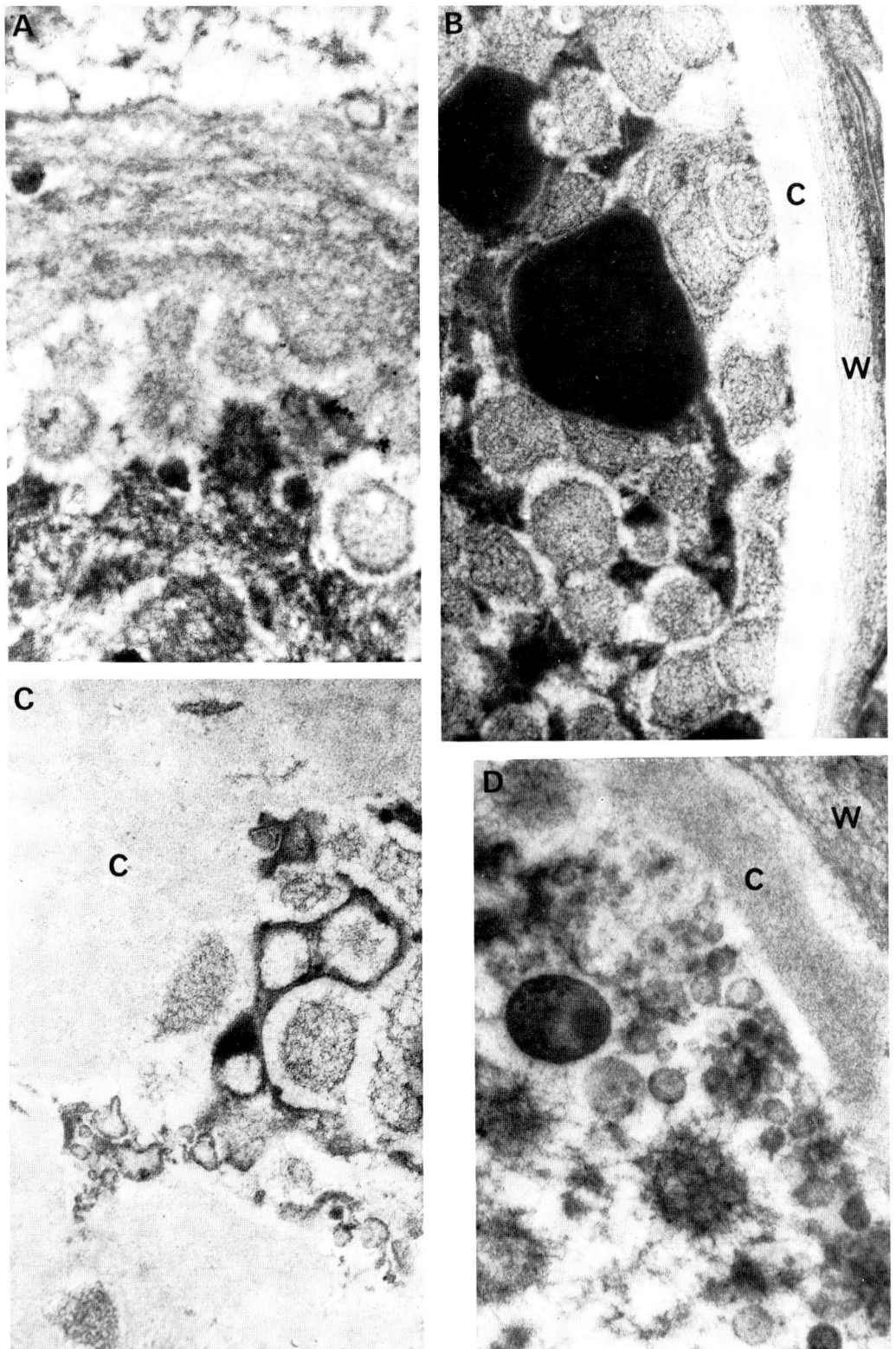

PLATE 13

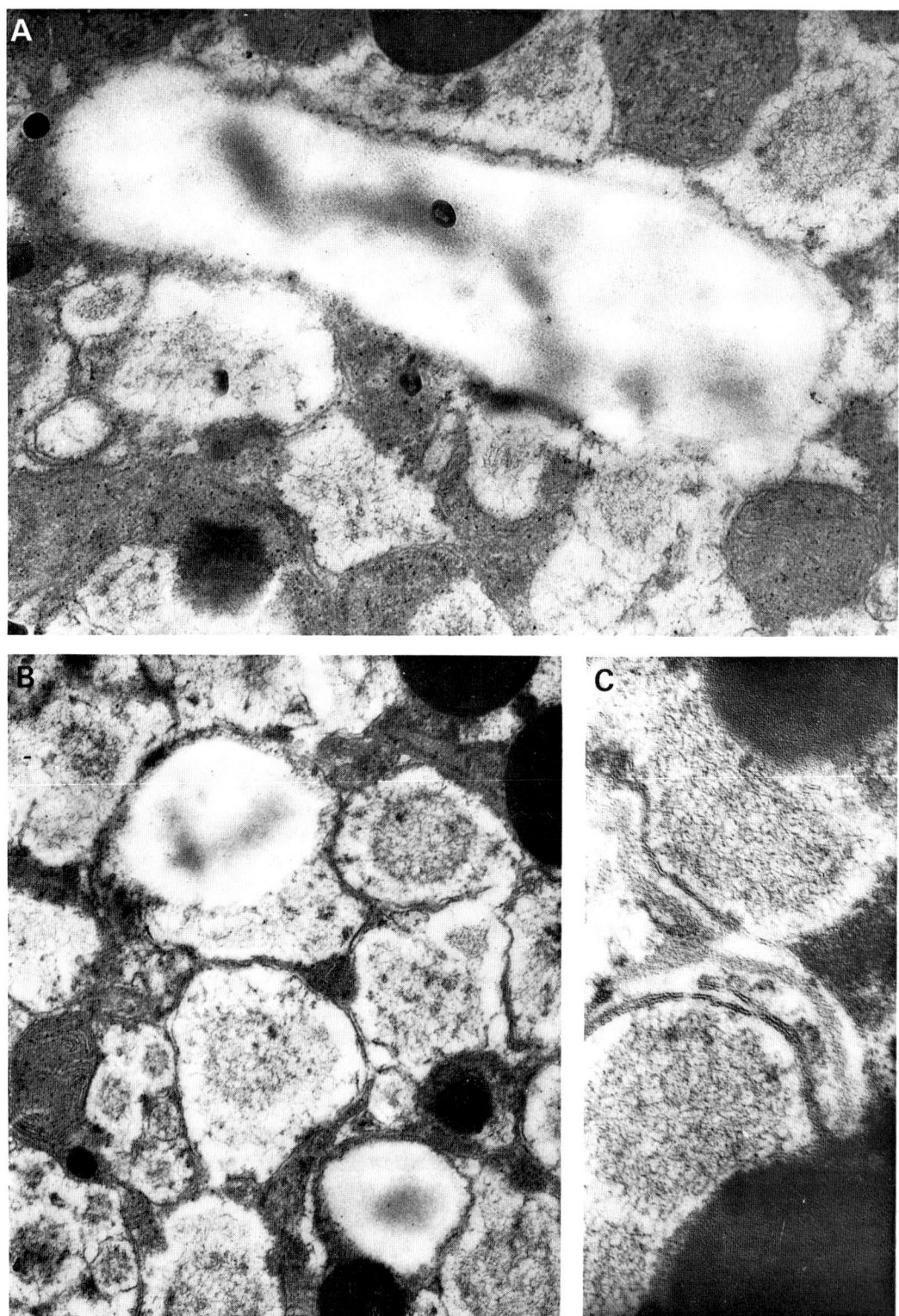

PLATE 14

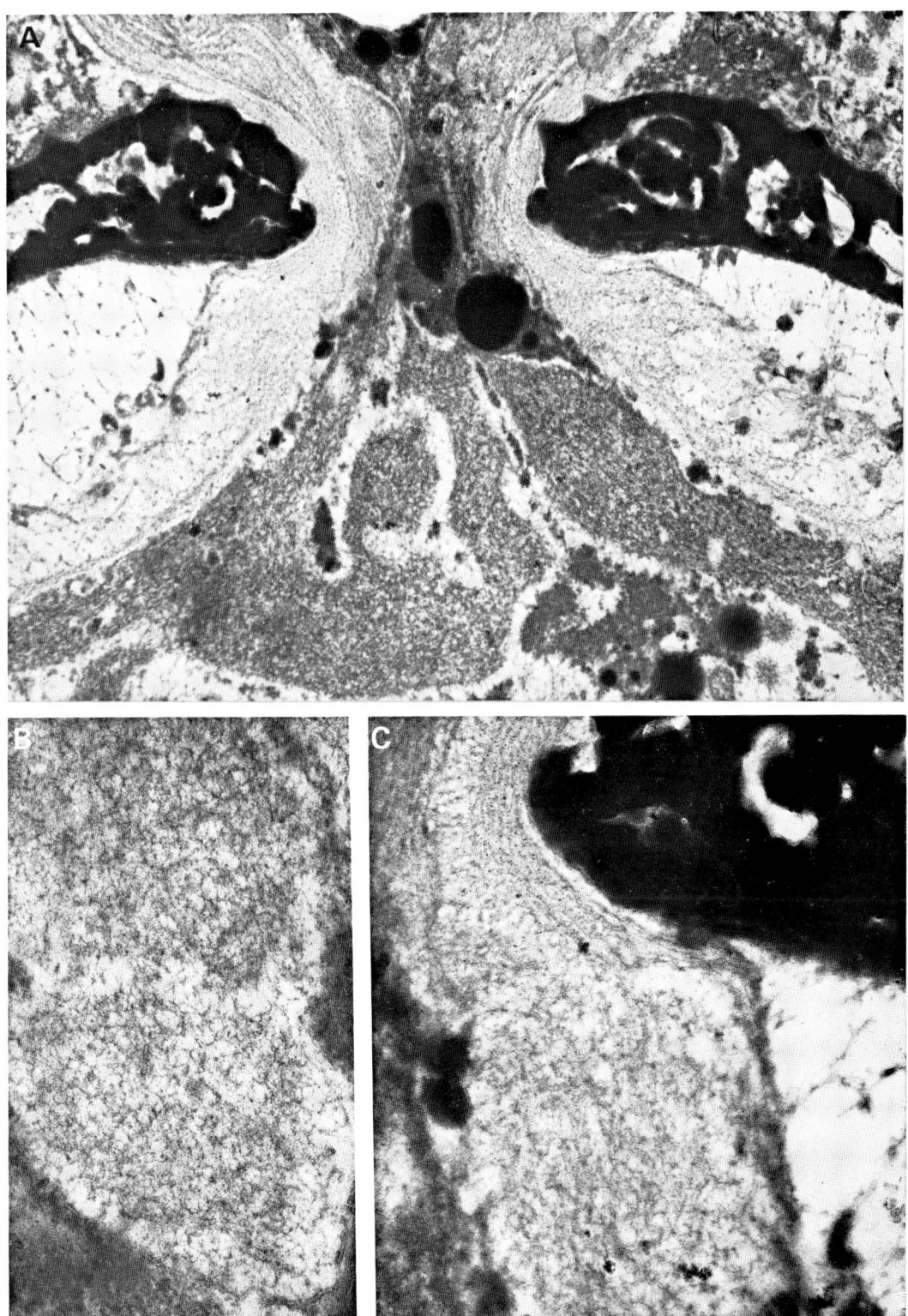

PLATE 15

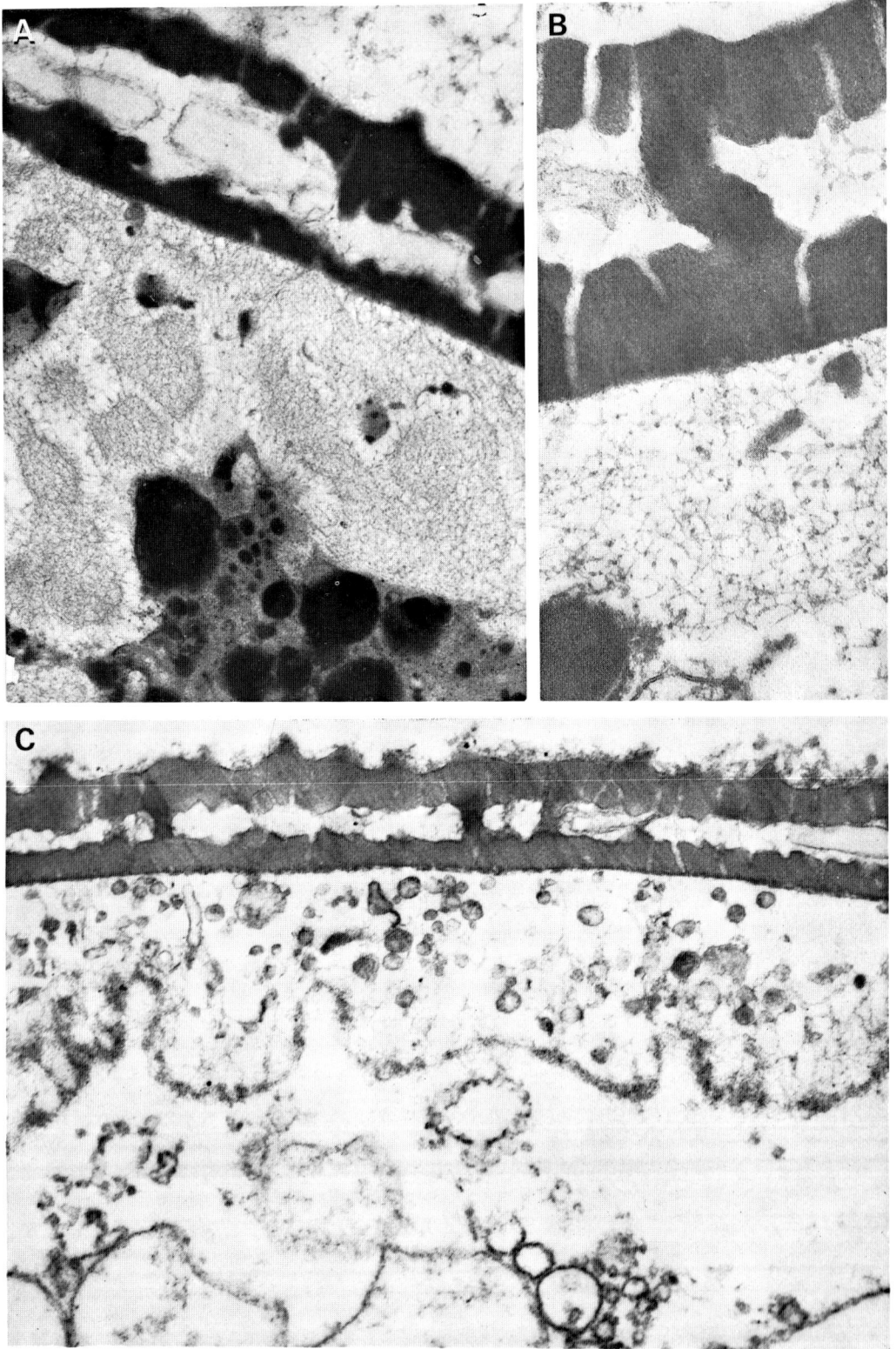

PLATE 16

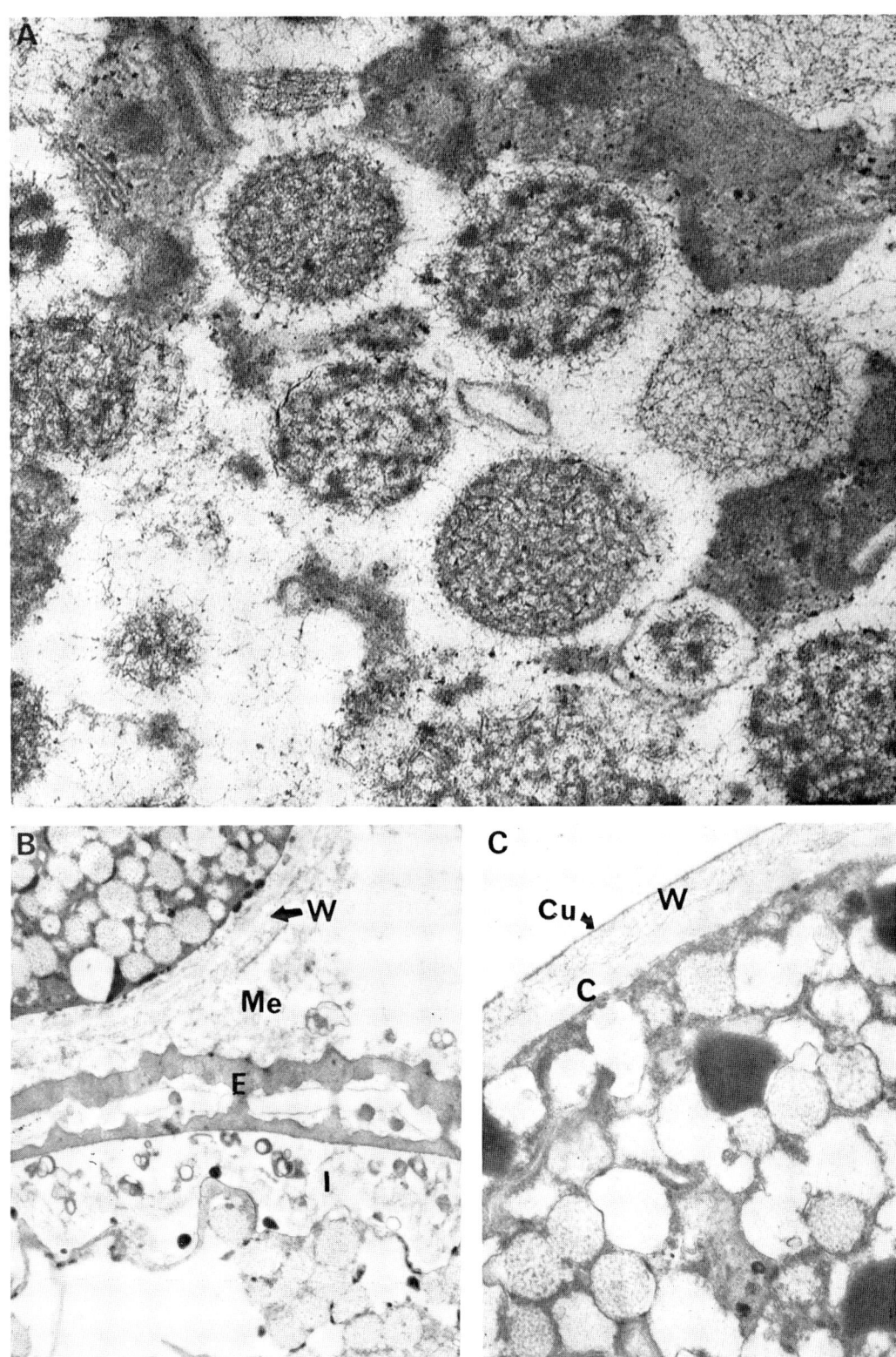

PLATE 17

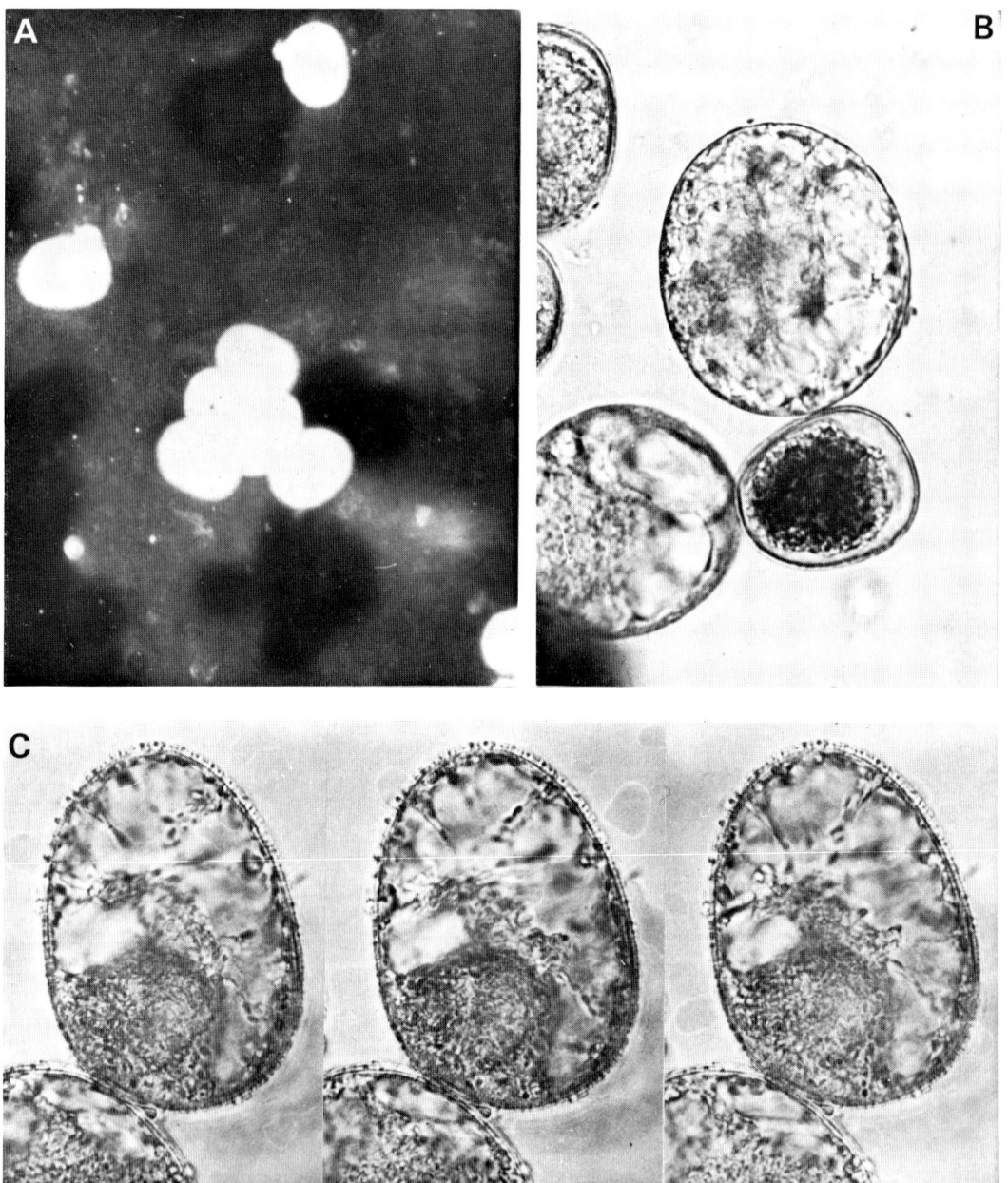

PLATE 18